The History of Science
Volume 7
Physical Science in the Nineteenth Century

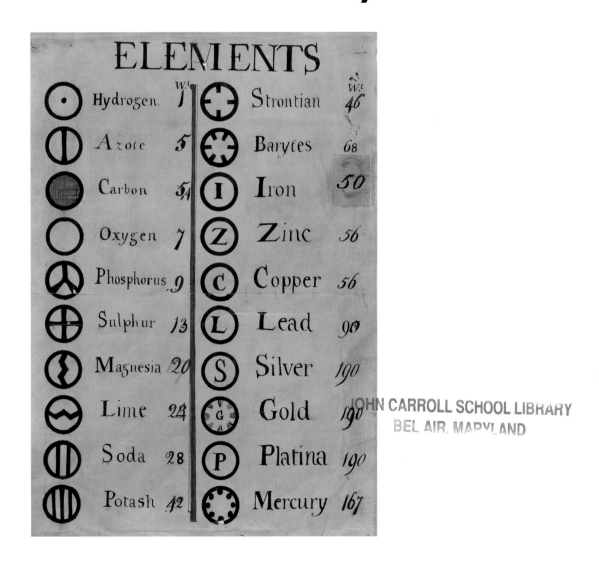

Dr. Peter Whitfield

First published in the United States in 2003 by Grolier Educational, a division of Scholastic Library Publishing, Sherman Turnpike, Danbury, CT 06816

For Compendium Publishing

Contributors: Sandra Forty

Editor: Felicity Glen

Picture research: Peter Whitfield and Simon Forty

Design: Frank Ainscough/Compendium Design

Artwork: Mark Franklin/Flatt Art

Reproduced by: P.T. Repro Multi Warna, Indonesia.

Printed in China by: Printworks Int. Ltd

Library of Congress Cataloging-in-Publication Data

Whitfield, Peter, Dr.

History of science / Peter Whitfield

    p. cm.

    Includes index.

    Contents: v. 1 Science in ancient civilizations – v. 2 Islamic and western medieval science – v. 3 Traditions of science outside Europe – v. 4 The European Renaissance – v. 5 The Scientific Revolution – v. 6 The eighteenth century – v. 7 Physical Science in the nineteenth century – v. 8 Biology and Geology in the nineteenth century – v. 9 Atoms and galaxies : modern physical science – v. 10 Twentieth-century life sciences.

    ISBN 0-7172-5729-0 (act : alk. paper) – ISBN 0-7172-5703-7 (v. 1 : alk paper) – ISBN 0-7172-5704-5 (v. 2 : alk paper) – ISBN 0-7172-5705-3 (v. 3 : alk paper) – ISBN 0-7172-5706-1 (v. 4 : alk paper) – ISBN 0-7172-5707-X (v. 5 : alk paper) – ISBN 0-7172-5708-8 (v. 6 : alk paper) – ISBN 0-7172-5709-6 (v. 7 : alk paper) – ISBN 0-7172-5710-X (v. 8 : alk paper) – ISBN 0-7172-5711-8 (v. 9 : alk paper) – ISBN 0-7172-5712-6 (v. 10 : alk paper)

    1. Science–History–Juvenile literature. [1. Science–History.] I. Grolier Educational (Firm) II. Title.

Q125 .W586 2003

509—dc21

2002029844

Acknowledgments

The publishers would like to thank the following for their help with the illustrations: Venita Paul and Sarah Sykes at the Science & Society Picture Library, Science Museum Exhibition Road, London SW7 2DD.

Picture credits

All maps and artwork are by Mark Franklin/Flatt Art.

All photographs were supplied by the Science & Society Picture Library except those on the following pages (T=Top; C=Center; B=Below): Author's collection p.5 (T), p.11, p.18 (T), p.25 (T), p.27 (T—British Library), p.42, p.50, p.54 (B), p.56 (B), p.64 (B), p.68; p.2 (Royal Institution/Science & Society Picture Library), p.16, p.17, p.52 (B—National Museum of Photography, Film and TV/Science & Society Picture Library); cover (National Railway Museum/Science & Society Picture Library).

Note

Underlined words in the text of this volume and other volumes in the set are explained in the Glossary on page 70.

Contents

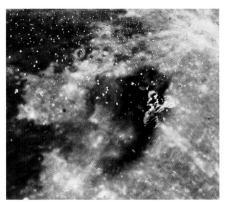

The Machine Age

Opposite, Above: The machine age
saw a major change in the way
that people perceived science. The
invention of machines such as this
beam engine transformed
manufacturing industry and
provided seemingly limitless power
in a way impossible to conceive in
earlier centuries.

Below: An electrical generator in
Berlin at the end of the nineteenth
century—electrical power was to
underpin twentieth-century
industry and social life.

The science of matter—the forms that matter can take and the forces that determine how it behaves—had baffled all the natural philosophers for centuries. It was during the nineteenth century that science began to move toward systematic answers to these questions, and some of the fundamental laws of physics and chemistry were formulated. These sciences were developed not on a purely theoretical level, but were based on experiment, above all perhaps on the study of machines. But the theoretical knowledge gained in this way was then used to build improved machines, and for the first time science and technology moved hand in hand. This combination began to revolutionize society and reshape the way people led their lives. The machine became the model through which physical forces could be understood and reduced to scientific laws, and those laws were in turn applied to the design of superior machines. In this sense the nineteenth century was "the machine age" not simply because it invented new machines, but because the machine influenced the way people understood the world.

KNOWLEDGE AND POWER

In Shakespeare's England Francis Bacon had proclaimed that knowledge was power, and he had foreseen a society transformed by scientific knowledge—by humans' control of natural forces. In its own day this prophecy rested on very slender foundations, and many generations would pass before it was fulfilled. Galileo, Newton, Descartes, Buffon, Laplace, and many others had opened new intellectual horizons that Bacon could never have dreamed of, but the material conditions of society remained unchanged.

It was the Industrial Revolution, between the years 1780 and 1830, that transformed science into a social force through the agency of technology. The machine changed millions of lives, it made science visible, and it proved the truth of the scientific principle that the world operates by laws and mechanisms that could be discovered and harnessed by humankind. In 1860 the English historian Lord Macaulay penned a famous eulogy on science, but it was really technology that he was describing: It was the machine that had "furnished new arms to the warrior … spanned great rivers and estuaries … lightened up the night with the splendour of the day … multiplied the power of human muscles … accelerated motion and annihilated distance … and enabled man to descend to the depths of the sea and soar into the air."

This revolution began with the steam engine, but later in the century electrical power and the chemical industry would bring about a second industrial revolution. This process of technological innovation acquired a momentum that has never ceased, and that seems, in fact, impossible to stop. It has created a culture of change, development, and improvement that now shapes all our lives. This culture is overwhelmingly based on technology, mechanical and electrical, and so is underpinned by the laws of physics and the science of matter.

Below: A coal-gas plant around 1850. The use of coal-gas for lighting had been pioneered as early as the 1790s. It played a central role in the industrial revolution, permitting factories to work through the night.

Heat and Energy

Above: James Joule, who formulated the principle of "the mechanical equivalent of heat." This is an 1882 painting by John Collier.

Below: Sadi Carnot, the French engineer who laid the foundations of thermodynamics.

The foundations of classical physics began with the question: What is actually happening in the operation of the steam engine? How, in the quaint words of Thomas Savery, could "water be raised by the agency of fire"? Several thinkers in the early nineteenth century were intrigued by this problem, and they recognized that underlying the question about the steam engine lay the operation of heat. Already in the 1780s the Scottish scientist Joseph Black (1728–99) had drawn a vital distinction between heat and temperature.

Black's experiments showed him that the heat required to produce a certain change in temperature was not the same for all bodies, and that different material therefore had different "specific heats." Heat must therefore be capable of interacting with matter to produce various effects. Perhaps it was even contained in matter as a subtle fluid, as Black and many of his contemporaries believed, that they named "caloric."

The most far-reaching study of heat as a principle of physics came from the pen of a French engineer, Sadi Carnot (1796–1831), who in 1821 published a short book entitled *Reflections on the Motive Power of Fire*.

Carnot's great insight was that what really happened in a steam engine was that heat was moved from a source to the components of the engine. That heat produced steam, which entered the cylinder under pressure and moved the piston. The net result was that work was done: The mechanical work performed by the steam engine arose from the transfer of heat, and heat could therefore be seen as equivalent to motive power.

But it is also true that the operation of motive power produces heat in both engines and animals. Therefore Carnot argued that the transformation between heat and work is theoretically reversible in an enclosed system. The heat will neither increase nor diminish, but remain constant, merely moving through the system in different forms. This theory was easier for Carnot to arrive at because he mistakenly believed in the old idea that heat was "caloric," a substance with real physical existence.

GETTING WARM

Wherever there exists a temperature difference, Carnot reasoned, there existed the possibility of generating motive power. He gave an analogy from the older technology of water power. Waterwheels were turned by the force of falling water. The difference in level between a head of water and its base creates the possibility of turning a wheel as it falls, and the greater the fall, the greater the potential power. So with heat, the greater the temperature difference, the greater the motive power. A high-pressure steam engine was more powerful than a low-pressure one.

Carnot observed that heat was one of the grand moving agents in the world of nature, powering winds and ocean currents, since they were now known to be caused by air or water moving from a higher to a lower temperature. Carnot's work laid the foundations of the science of <u>thermodynamics</u> (from two Greek words meaning "heat-power"), but sadly he died too young to see the fruition of his ideas.

WIDER APPLICATION

A similar approach to Carnot's was taken later in England by James Joule (1818–89). Joule designed a series of experiments to measure precisely how much work was needed to perform simple mechanical tasks, such as turning a small wheel. Conversely, he also measured how much heat was generated in and around the mechanism. Joule discovered that there was always an equivalence between heat and work. Joule focused on a simple unit of heat—that needed to raise the temperature of one pound of water by one Fahrenheit degree—and a simple unit of work—that needed to raise a one-pound weight one foot. After lengthy experiments he found that the "mechanical equivalent of heat" was 838 foot-pounds. This means that if a small winch is turned, by hand, for example, in a pound of water so as to raise 838 one-pound weights, then heat will be generated that will raise the temperature of the water by one degree (see diagram below).

But this was not all, for Joule found the same equivalence in chemical and electrical reactions. If zinc is plunged into concentrated acid, it dissolves quickly and generates sudden heat. But if the zinc is made the element in a battery, it will heat a wire in a more controlled way. If that current is made to drive a small electric motor, then the heat in the wire is less. In each case the amount of heat varies in proportion to the work done. Joule called work in this context "living force," stating that "wherever living force is apparently destroyed, an exact equivalent of heat is restored; the converse is also true, namely that heat cannot be lessened or absorbed without the production of living force … in these conversions nothing is ever lost." Joule's painstaking experiments built on the work of Carnot and prepared the ground for the mature statement of the laws of thermodynamics, and his name is preserved in the <u>joule</u>—the basic unit of energy used in the scientific world.

JAMES JOULE (1818–89)
- Natural philosopher.
- Born Salford, England.
- Educated by private tutors and went on to study chemistry in particular.
- 1840 Discovers the Joule Effect—heat produced in a wire by electric current is proportional to the resistance and to the square of the current.
- 1843–78 Through experiments showed that heat is a form of energy determined by the amount of energy applied.
- 1853–62 Worked with Lord Kelvin (William Thomson) on temperature change in gases, the "porous plug" experiments—showed that when a gas expands without doing external work, its temperature falls; known as the Joule-Thomson effect.
- Formulated the absolute scale of temperature.
- First to describe magnetostriction—the change in size of ferromagnetic materials when placed in a magnetic field.
- An SI unit of work, quantity of heat, or energy, a joule, is named after him.

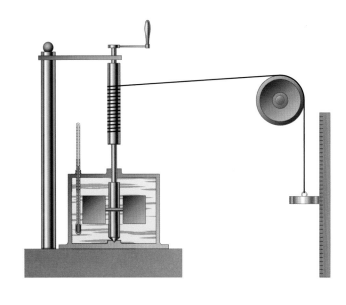

Joule's apparatus for measuring the mechanical equivalent of heat. By winding the winch attached to a weight, a paddle wheel is turned in a bowl of water. The heat generated by this work is measured by a thermometer.

The Science of Thermodynamics

Above: Hermann von Helmholtz.

Opposite: A thermodynamic cycle: Heat is converted to work, which can be stored or reconverted to heat; theoretically this cycle could go on for ever, but in practice some of the heat will be lost into the apparatus and the environment, and the cycle must end.

Below: Rudolf Clausius.

The all-important study of heat power—thermodynamics—was pioneered mainly by a group of German scientists, but the Englishman William Thomson, later Lord Kelvin, also played a leading role. It was in the 1850s that Kelvin and the German physicist Rudolf Clausius, working independently, articulated the laws of thermodynamics. Carnot and Joule had both expressed what would later become known as the first law—that heat in an enclosed system is conserved. Clausius, in his work *On the Motive Force of Heat* (1850), spelled out the second law—that heat does not flow spontaneously from a colder to a hotter body, but always the reverse. Heat can be made to flow from colder to hotter, but only through the intervention of work from outside—this is the principle on which refrigeration operates.

Kelvin's name has become identified with the discovery that there is a point at which it becomes impossible to remove any more heat from a body because to do so one would have to provide a temperature difference by creating yet lower temperature, which is impossible. Absolute zero on the Kelvin scale is −273°C (degrees Centigrade—equivalent to −459°F), which Kelvin derived theoretically, not experimentally. The impossibility of taking heat from an object without limit became the third law of thermodynamics.

UNIVERSAL LAW

Clausius went further and enlarged the second law by introducing the vital concept of "<u>entropy</u>" (from the Greek word meaning "transformation"), meaning the decreasing availability of heat to do work. Any machine or any system will steadily use up available heat; and although that heat is not destroyed, it is dissipated or slowly lost and can only be recovered by more work, requiring more heat input. This second law of thermodynamics is one of the fundamental laws of the universe. It gives a scientific expression to what we all know: that things will naturally run down, weaken, decay, and die unless they are renewed, and often even then.

This may seem to contradict the first law and the argument that Carnot had put forward: that in any ideal system the heat cycle is reversible. The reversibility principle is true, but it is true in theory only; in practice some heat is always dissipated in ways that make it irrecoverable. To give a simple example, gas is burned in moving a car. That burning accomplishes work, but in the process the engine, the transmission, and the tires get hot, and this heat is dissipated. At the end of the journey that heat still exists somewhere in the environment, but it cannot be directly recaptured and made to do more work. Hence the two laws do not in fact contradict each other.

It is basic to the distinction between <u>mechanics</u> and thermodynamics that according to mechanics, all the energy of a system is available to do work, while according to thermodynamics, only a part of the energy is so available. It is always the aim of the engineer to make any system as efficient as possible, that is, to utilize the maximum energy and not to waste it. However, the law of entropy explains why the idea of a perpetual-motion machine is an impossibility: No system can ultimately be self-sustaining, but must draw energy from outside or cease to function.

ENERGY CONSERVATION

Kelvin, Clausius, and another German scientist, Hermann von Helmholtz (1821–94), synthesized this thinking in developing the language of "energy," which not only replaced the old idea of caloric, but unified all ideas about physical processes, mechanical, chemical, and electrical, as Joule had shown. In particular, Helmholtz introduced the phrase "conservation of energy" as a link between all these types of physical operation. But when entropy was applied to the universe as a whole, Clausius and Helmholtz saw its profound implications: that the cosmos would eventually end with a "heat-death" when no more energy was available anywhere in the system. "The energy of the universe is constant," wrote Clausius, "but the entropy of the universe is always tending toward the maximum." In other words, the entire universe can be seen as a thermo-dynamic machine in which the sum of all energies remains constant through repeated transformations. Any form of physical activity was convertible to any other form—heat, work, chemical reaction, radiation, electricity.

Helmholtz was also a biologist, and he applied the same model to animal physiology, showing that muscular action involves the oxidization of sugar, producing energy. With the contemporary understanding of photosynthesis it was evident that plant growth too was a way of recycling solar

RUDOLF CLAUSIUS (1822–88)
- Physicist.
- Born in Köslin, Germany.
- Attended the University of Berlin, initially to study history, but changed to science.
- Started teaching physics at the Royal Artillery and Engineering School, Berlin, 1850; moved on to teach at Zurich, 1855; Würzburg, 1867; and Bonn, 1869.
- 1850 At the same time as Lord Kelvin he devised the second law of thermodynamics—that heat cannot pass from a colder body to a hotter one.
- 1858 Introduced the concepts of mean free path and effective radius.
- Worked on optics and electricity.
- 1865 Introduced the word "entropy" to describe that dissipation is equivalent to entropy increase.
- 1869 Became professor of natural philosophy at Bonn.
- Among other areas of interest he investigated electrolysis and calculated the mean speed of gas molecules.
- He was influential in establishing thermodynamics as a science

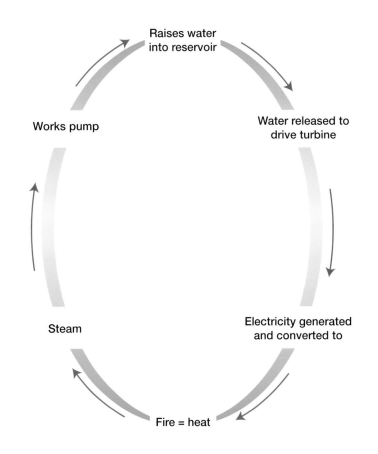

9

WILLIAM THOMSON, LORD KELVIN (1824–1907)

- Mathematician and physicist.
- Born in Belfast, Northern Ireland.
- 1832 Moved with his family to Glasgow.
- Aged 16 went to Peterhouse, Cambridge University; after graduating went to Paris to study.
- 1842 Published a paper showing how to solve problems in electrostatics.
- 1846–99 Professor of mathematics and natural philosophy at Glasgow; became interested in physics, uniquely combining pure and applied science.
- 1848 Helped devise the absolute temperature scale—now given in degrees Kelvin.
- 1850 At the same time as Rudolf Clausius, he devised the second law of thermodynamics.
- Worked particularly in the areas of hydrodynamics, especially wave motion and vortex motion: appointed chief consultant for the laying along the ocean bed of the transatlantic cable, 1857–58.
- Researched in the area of geomagnetism; invented many electrical instruments; improved ships' compasses.
- Patented a mirror galvanometer for speeding telegraphic transmission, making him a wealthy man.
- Many of his inventions were made by his own company, Kelvin & White.
- 1892 Created First Baron Kelvin of Largs.

energy. Dramatic new horizons were being opened by the concept of energy conservation, so that the physical and the living realms of the universe might be seen as bound together in a complex, interdependent system. Newton's theory of gravitation had seemed to explain something fundamental about the large-scale structure of the universe, and now thermodynamics seemed to reveal something equally profound about its inmost workings. Heat had become a dynamic force, while "caloric" was a static substance. The nineteenth-century view of the importance of the machine could receive no higher expression than this—that the whole universe could be seen as a machine functioning according to the laws of physics.

SCIENCE OF THE UNIVERSE

Helmholtz and his contemporaries were missing some vital pieces of the jigsaw because they did not yet know how forces operated at the atomic level, nor did they know the true nature of the energy that fires the Sun and stars, although they guessed that this played a central role in the cosmic system. Nevertheless, Helmholtz did build the ideas of thermodynamics into an intriguing model of scientific explanation. If we imagine, he said, all matter split into its elementary particles (whatever they may be), then all conceivable changes, all the forms that nature takes, are reducible to the spatial

Lord Kelvin who, with Helmholtz, developed the language of energy, entropy, and thermodynamics. He is seen with his compass—an improvement on the existing compass mounted in a binnacle fitted with magnets and spheres.

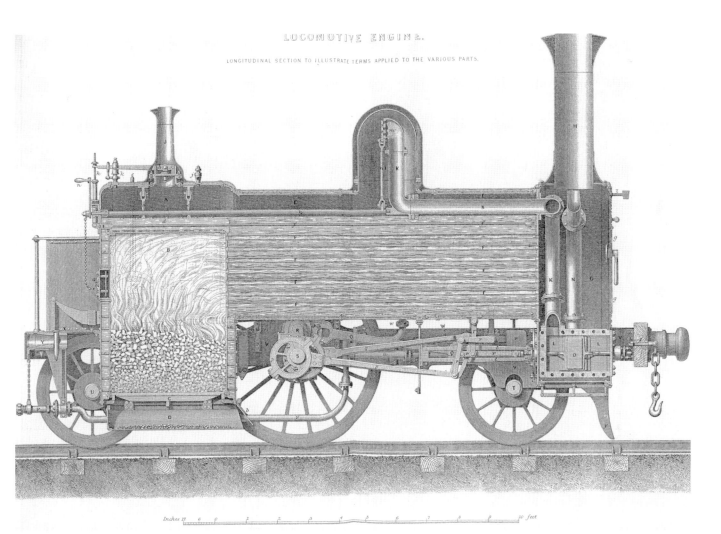

LOCOMOTIVE ENGINE.

LONGITUDINAL SECTION TO ILLUSTRATE TERMS APPLIED TO THE VARIOUS PARTS.

rearrangement of those elements. Therefore the whole realm of scientific investigation must focus on the forces that cause those changes, on identifying and measuring their energies, and tying them into a unified system. In this argument classical physics is, again, the fundamental science of the universe.

The steam engine whose workings led physicists to an understanding of thermodynamics.

IMAGINING THE ATOM

A generation after Helmholtz and Clausius two other great scientists, the Austrian Ludwig Boltzmann (1844–1906) and the American Josiah Gibbs (1839–1903), introduced atom theory into the discussion of thermodynamics. They showed that heat and entropy could be considered as arising from the behavior of a large number of atoms. In this approach heat began with the movement of atoms, and the effects of these movements could be analyzed mathematically. They produced formulas for, for example, the way that the atoms in gases would behave when the gas was heated or cooled: how they absorbed or conserved energy. This work was to be increasingly important in later atomic theory (see Volume 9), but scientists of this time had no knowledge of what the atom really was: It was a working hypothesis, not a physical reality, so this work opened a new field known as statistical thermodynamics because it dealt with mathematical probabilities rather than physical models.

Electricity and Magnetism

Above: Oersted discovered the connection between electricity and magnetism with an apparatus: The compass needle was deflected when a current flowed around it. This is the paper that announced his discovery.

Opposite, from Top to Bottom: Ampère, Ohm, and Oersted.

Below: Oersted's needle of 1828.

The other field in which classical physics was taking shape was in the investigation of electricity. The importance of this study lay in its proof that the physical world did not consist of matter alone but also of forces; indeed, that matter was not inert but was apparently held together by forces whose power people could begin to unlock and manipulate.

In the 1820s the German scientist Georg Ohm (1787–1854) began to investigate what happened physically when an electric current passed through a wire. He experimented with wires of different materials, different thicknesses, and different lengths, using Coulomb's torsion balance (see Volume 6, page 36) to measure the strength of the current. He was able to show that the thicker the wire, the weaker was the current that flowed from any given battery, and he found that what he called the "resistance" emerged as heat. He suggested that electricity was flowing from particle to particle within the wire and was caused by electrical "tension," or a "potential difference" that existed between the two ends of the circuit. It would later be termed "electromotive force" because it was able to do work—it was in fact a form of energy exactly in line with the findings of the new study of thermodynamics. Ohm's name became attached to the unit of resistance: the amount of electromotive force that is lost as it is carried through any medium. Resistance is what causes current electricity to become heat in an electric coil.

THE SAME INVISIBLE FORCE

At exactly the same time as Ohm, the Danish physicist Hans Christian Oersted (1777–1851) discovered—apparently by accident while he was delivering a lecture—the close connection between electricity and magnetism. He observed that a wire carrying an electric current would deflect the needle of a nearby compass. Oersted concluded immediately, and correctly, that electricity and magnetism were two types of the same invisible natural force. His discovery aroused much excitement in the scientific world, and his experiments were widely repeated.

The most systematic work in this field was carried out in Paris by André-Marie Ampère (1775–1836) in a series of experiments during the 1820s. Ampère showed that a current deflects a magnet's needle if they are parallel, but not if they are at right angles. Moreover, the direction in which the needle turns depends on the direction in which the current flows. He then showed that electrified wires themselves act as magnets: They attract each other when their currents are flowing in the same direction, but repel each other when their currents are opposed. Finally, he demonstrated that a cylindrical coil of wire when electrified behaves exactly like a magnet whose attractive power increases as the current becomes stronger. This was the electromagnet, and within a few years the American physicist Joseph Henry had constructed an electromagnet that could lift a ton weight. The electrical deflection of the magnetic needle led to the first galvanometer, invented by Ampère but named in honor of Galvani, which measures the strength of an electric current by the amount by which it deflects the needle.

Ampère concluded from his experiments that a magnet is a body in which an electric current is permanently circulating, while other bodies can be electrified temporarily. "A magnet is only a permanent collection of electric currents," he wrote. He calculated that electromagnetic effects obey an inverse-square law, just like Newton's law of gravitation: The attractive force is the exact opposite of the distance between the magnet and its object. Ampère's name is preserved in the <u>amp</u>, the basic measure of the strength of an electric current.

DAVY'S CONTRIBUTION

Between them Ohm, Oersted, and Ampère had raised the study of electricity to a new level far beyond the parlor games typical of the eighteenth century. The availability of current electricity from a battery made it much more accessible to investigation compared with the static electricity that earlier experimenters had used. A further insight into the functions of electricity was opened up by Sir Humphry Davy (1778–1829), who proved by experiment that an electric current could have chemical effects. Using a battery, Davy was the first to isolate sodium and potassium, which are elements so reactive that they do not occur in nature in a pure form. It was evident that electricity was a force that was somehow linked to the laws of chemistry, thermodynamics, and gravity, but its nature still remained highly elusive. Other forces such as gravity or air pressure were constantly present, but how could scientists explain this mysterious force, which nature could apparently switch on or off in different materials under different conditions?

Electricity:
Faraday, Maxwell, and Hertz

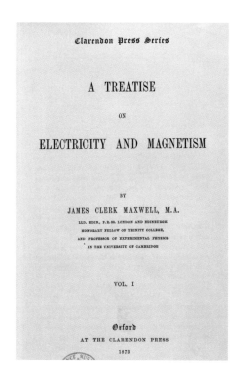

**MICHAEL FARADAY
(1791–1867)**
- Chemist, physicist, and creator of classical field theory.
- Born in Newington Butts, England.
- Apprenticed to a bookbinder; reading the books, he developed an interest in science.
- 1813 Became Humphry Davy's assistant for a trip to Europe.
- 1827 Elected to Davy's chair of chemistry at the Royal Institution.
- 1827 Published *Chemical Manipulation*.
- 1839–55 In *Philosophical Transactions* he published a series of articles called "Experimental Researches on Electricity."
- 1845 Described the rotation of polarized light by magnetism—the Faraday effect.
- 1862 Became adviser to Trinity House (the body that looks after British coastal safety) and advised the use of electric lights in lighthouses.

Continuing experiments in the 1820s and 1830s revealed still further implications of the electrical force. The leading figure in this field was Michael Faraday (1791–1867), whose life is of great interest in showing how scientific knowledge could transcend social barriers. Faraday was a young man of humble birth and virtually no formal education who became an assistant to Sir Humphry Davy (see page 13) at the Royal Society in London, a center of research and public lectures in the sciences, and who succeeded in becoming its leading experimenter and lecturer. Physics was still in its infancy, with no textbooks to study and no laws to master, so that Faraday was able, within a few short years, to place himself at the frontiers of science. He was a born scientist, with an innate sense of how natural forces work, and how experiments should be designed. His one weakness was his lack of mathematical training, which meant that he could not analyze and quantify his results, a task that had to be left to others.

ELECTRICITY AS A POWER SOURCE

The first area in which Faraday's experiments were so important was in electrolysis. Building on the work of Humphry Davy, he showed that when an electric current was passed through a solution

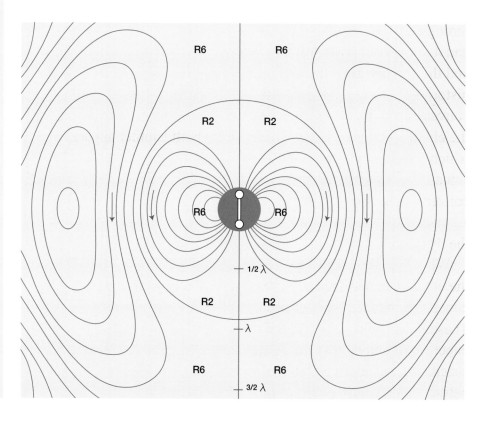

containing chemical salts, the salts were decomposed into their constituent elements. An example would be a copper sulfate solution in which copper and sulfur were released, or water could be resolved into its constituents, hydrogen and oxygen. The amount of decomposition was directly proportional to the duration and strength of the current: Five minutes' current would release exactly half as much as ten minutes' current.

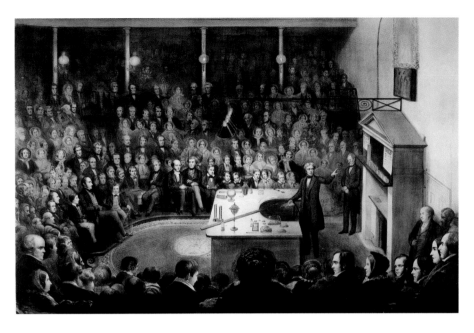

The implication of this was that the electrical charge was an integral part of matter and was involved with the way matter was bonded together, so that the arrival of a new charge negated or dissolved the original bond. Today we understand that this happens at the atomic level, but the language of chemistry had not evolved in Faraday's time (see below, pages 22–28). These experiment with electrolysis were of great long-term importance for the science of matter, but more obvious practical implications could be found in his demonstration that electricity was a potential power source.

MECHANICAL ENERGY

In 1821 Faraday designed an apparatus in which a copper rod was suspended near a magnet. When an electric current was passed through the rod, it rotated around the magnet. Faraday's apparatus had a second mirror-image component in which the rod was fixed and the magnet was free, and in this the magnet rotated. This historic experiment showed that electrical energy could be converted into mechanical energy, and that is the basic principle of the electric motor. Faraday subsequently built another apparatus in which a magnet was moved through a wire coil. While it was actually moving, it caused an electric current in the coil. Again the converse was also true: A copper disk rotating between the poles of a horseshoe magnet created a current, and that became the basis of the dynamo.

When he asked himself how these effects were produced, how the electromagnetic force operated across empty air, Faraday conceived the idea of an electrical field that set up lines of force. These lines could in fact be traced by simply placing iron filings on paper close to the charge and observing their patterns. Faraday speculated that electricity functioned by changing the relationship between elementary particles of matter—whatever they might be.

Above: Faraday lecturing on electricity before Prince Albert, Queen Victoria of Great Britain's consort, at the Royal Institution, December 27, 1855.

Opposite, Above: Title page of Maxwell's *Treatise on Electricity and Magnetism*, 1873.

Opposite, Below: Hertz's diagram of radio waves as published in *Ausbreitung der elektrischer Kraft* ("The Propagation of Electrical Force") of 1892.

Below: Maxwell's picture of the lines of force in an electromagnetic

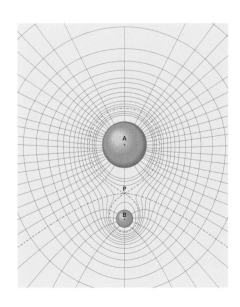

This is quite close to the modern understanding of electricity as a flow of electron jumps.

MATHEMATICAL STRUCTURE

Faraday was not able to give his work any final theoretical or mathematical expression. That achievement fell to James Clerk Maxwell (1831–79), who set out to show that there was a strict mathematical structure to the forces with which Faraday had experimented. Essentially, Maxwell built a mechanical model to describe how electrical and magnetic forces reproduce themselves in space. He accepted Faraday's concept of the electrical fields and showed their strengths and patterns. His central conclusion was that these fields consisted of high-velocity waves. When he calculated the characteristics of these waves, he was astonished to discover that they were essentially the same as light waves and traveled at the same speed. Maxwell had discovered "electromagnetic radiation," and that light was a form of this radiation. He predicted that other forces with different wavelengths would be revealed within a spectrum.

RIGHT WAVELENGTH

As early as 1800 the German-born British astronomer William Herschel (1738–1822) had noticed, while studying the spectrum of sunlight, that heat was clearly measurable beyond the red end of the spectrum, that is, outside the limit of visible light. Herschel had correctly guessed that some form of solar radiation, invisible to the

Above: An early electromagnet built by Faraday in the 1820s.

Below: Maxwell was the first to understand that electrical fields behaved like high-velocity waves in exactly the same manner as light. Scientists later confirmed the existence of a spectrum of electromagnetic radiation, of which visible light is but one small part.

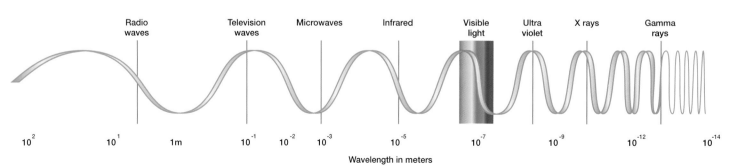

Radio waves   Television waves   Microwaves   Infrared   Visible light   Ultra violet   X rays   Gamma rays

10^2 10^1 1m 10^{-1} 10^{-2} 10^{-3} 10^{-5} 10^{-7} 10^{-9} 10^{-12} 10^{-14}

Wavelength in meters

Faraday at work in his chemical-physical laboratory.

human eye, was responsible, and indeed it would later be identified as infrared radiation. In 1868 the Swedish physicist Anders Angstrom (1814–74) produced the first map of the solar spectrum and devised a way of estimating the wavelengths of the different forms of radiation. His unit of one ten-millionth of a millimeter was later called the angstrom, the smallest unit of measurement in scientific use.

Seeking experimental proof of Maxwell's models, Heinrich Hertz (1857–94) succeeded in generating electromagnetic waves in a laboratory in Karlsruhe, Germany, in the 1880s. From an unclosed electrical circuit Hertz projected a current that he was able to detect across space by means of an unclosed loop of wire. He refracted these waves through a prism and with mirrors, just as light is bent, and estimated the length of the waves. From his results we now know that these were radio waves, moving as light moves, but beyond the visible spectrum. The discovery of the <u>electromagnetic spectrum</u> was another revelation of unity in the forces of nature, but one whose implications were even more puzzling than those of thermodynamics.

HEINRICH HERTZ (1857–94)
- Physicist.
- Born in Hamburg, Germany.
- Schooled at the Johanneum Gymnasium before moving on to Berlin to study.
- Became assistant to Hermann von Helmholtz.
- 1885 Appointed professor of physics at Karlsruhe University.
- 1889 Moved to Bonn University.
- 1887 Discovered "Herzian waves" (radio waves).
- Main researches were in the field of theoretical thermodynamics, especially working on electric waves.
- The SI unit of radio frequency is named for him and defined as the number of complete cycles per second.

The Problem of the Ether

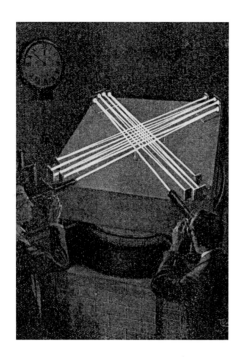

Above: The Michelson-Morley apparatus, which attempted to detect the movement of the Earth through the ether—and failed.

Below: Light waves intermingling from a dual source: scientists asked what medium carried these waves—was it the ether?

One of the most baffling problems thrown up by the discovery of electromagnetic radiation was the means by which these forces traveled through space. The classical theory, supported by Newton, was that light was a stream of particles. But Christian Huygens had disagreed, proposing that light should be understood as a series of waves. In the early years of the nineteenth century scientists such as Thomas Young (1773–1829) in England and Augustin Fresnel (1788–1827) in France had taken up and developed the wave theory of light and made it widely accepted. Maxwell had extended the idea of waves to the other parts of the electromagnetic spectrum and had imagined these forces as waves rippling through space.

But the question was: A wave in what? Waves were familiar in water, and it was now generally understood that sound traveled through the air in waves. If light and electricity were waves, they must be waves in some medium. It could not be air, since light traveled from the Sun across airless space. It was easy to prove that neither light nor magnetism was affected by a vacuum; so what was the medium involved?

EXAMINING ETHER

Perhaps an answer lay in the ancient doctrine of the ether, the subtle, invisible substance that was believed to fill the universe. Newton had accepted it, and it had been essential to Descartes's concept of the vortices (see Volume 5, page 11). If it existed, the ether must have extraordinary properties: It must be so tough and elastic that light and

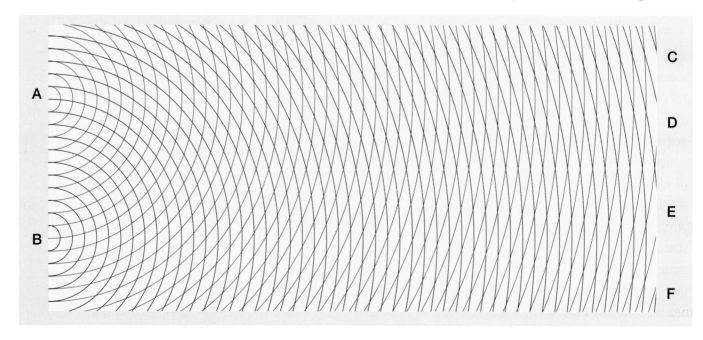

radiation could speed through it at an unimaginable velocity, yet so insubstantial that physical bodies such as the planets could whirl through it without effect. Above all, the ether was completely undetectable by any physical test, and it apparently continued to exist in a vacuum created in a laboratory. The leading physicists of the age—Kelvin, Maxwell, Helmholtz, and others— thought that all forces were fundamentally material in motion, and they therefore continued to uphold the concept of the ether as a kind of invisible sea in which the Earth was floating. Yet so elusive and contradictory were the qualities of the ether that some scientists doubted its very existence.

In an attempt to settle this question, two American physicists, A.A. Michelson (1852–1931) and E.W. Morley (1838–1923) set up a carefully structured experiment in Cleveland in the 1880s. Michelson and Morley reasoned that if light was indeed traveling through a medium, some difference ought to be detectable in the speed of light when it moved in the same direction as the Earth in its orbit compared with its travel across the orbit at right angles to the Earth. The Earth's orbital velocity through the ether—more than 60,000 miles per hour—is low compared with that of light, but surely high enough, they believed, to register a difference.

The two experimenters designed a sensitive apparatus in which a single beam of light was split into two, the first at right angles to the second. One part of the beam was aligned with the path of the Earth in space, the other crossing that path. Michelson and Morley then began to search for any measurable difference in the velocity of the two beams, but no difference could be found. It was all but impossible to conceive of a medium through which a body could move at a velocity of 60,000 miles per hour without producing any measurable effect. By the 1890s, when this experiment was reported throughout the scientific community, many came to believe that the ether was an illusion. In scientific theory it had been seen as a logical necessity, but in truth nature behaved as though it simply did not exist. The new physics of the following generation would finally show that the ether belonged among the scientific myths of the past, like caloric or phlogiston.

Albert Michelson, who set out to test the reality of the ether after becoming professor of physics at Cleveland, Ohio, in 1882. There he joined with the professor of chemistry, Edward Morley, to prove its nonexistence. Michelson's work would see him become, in 1907, the first American to be awarded the Nobel Prize. He is seen here in a photograph dated 1928.

Armand Fizeau Measures the Speed of Light

In all the nineteenth-century discussions about light, radiation, and energy, experiment was crucial. Scientists were no longer content to argue about concepts and theories, but sought to measure forces and effects in an attempt to build a new type of physics.

The transmission of light was one of the mysteries of nature, and scientists sensed that its almost inconceivable speed must be one of the great defining boundaries of physics. Estimates of that speed had been made from the late seventeenth century onward by astronomers who observed slight differences in the timing of celestial events (see Volume 5, pages 21–22), but it was typical of mid-nineteenth-century science to try to measure it more exactly. Technical advances in metalworking meant that in this period it was possible to make more finely calibrated instruments than ever before, which in turn encouraged scientists to devise ever more precise experiments. In 1849 the Parisian physicist Armand Fizeau (1819–96) conceived and built a mechanism (see illustration below) that he believed would measure the speed of light more accurately than before.

Armand Fizeau.

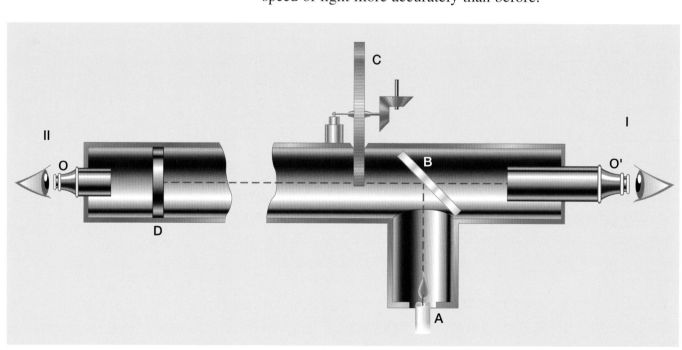

ALL DONE BY MIRRORS

The central component of Fizeau's apparatus was a toothed wheel (C) that would interrupt a beam of light (see illustration at right). This device was set up on a hilltop near Paris five miles away from an observer who trained his telescope on it. The beam of light was reflected in a mirror so as to double the distance that it must travel, allowing more time for the effect. With the light source and the observer—Fizeau himself—in place, the toothed wheel was rotated slowly at first, then fast and faster. What Fizeau was looking for was the moment when the teeth blocked the beam of light completely and the whirling teeth appeared to become solid. At that moment the time that one tooth took to follow another was less than the time taken by the beam of light to travel 10 miles.

Since Fizeau knew the circumference of the toothed wheel and the speed at which it was rotating, he could calculate what that tiny fraction of time was and say that in that time light had traveled ten miles. Using this method, Fizeau was able to calculate the speed of light as 195,000 miles per second, around 5 percent higher than the true figure obtained much later in more sophisticated experiments.

SCIENCE OF FORCES

Fizeau carried out other experiments into the nature of light, some of which helped the growing understanding of shifting wavelengths in starlight, and others that attempted to detect the existence of the ether, without success. His work was typical of the robust, innovative approach to experimentation among the scientists of his era, who were keenly aware that they were reshaping people's understanding of nature.

With the work of Faraday, Maxwell, Kelvin, and Clausius, the science of matter had become a science of forces: Heat, energy, electricity, and radiation had turned out to pervade the physical world. The explanation of these forces made up the foundations of classical physics. At what level these forces acted on physical matter was still an undiscovered secret that would begin to be understood in the atomic physics of the next generation.

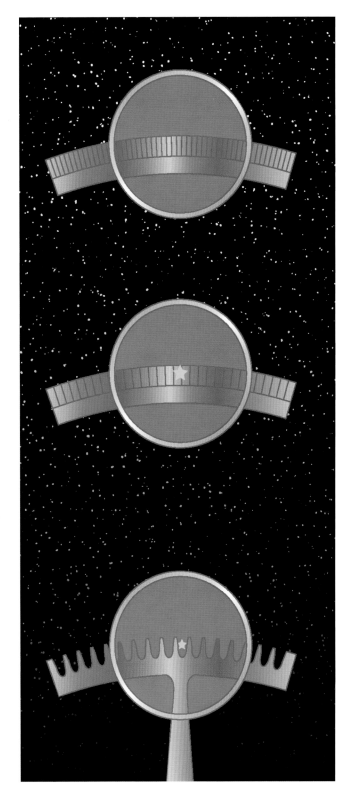

Above: As the toothed wheel rotated faster and faster, the light apparently vanished.

Opposite, Below: Fizeau's mechanism. The telescope on the left was sited five miles away from the toothed wheel and the light source.

The Beginnings of Modern Chemistry

**JOSEPH PRIESTLEY
(1733–1804)**
- Clergyman and chemist.
- Born in Leeds, England.
- 1755 Became a Presbyterian minister at Needham Market; wrote *The Scripture Doctrine of Remission.*
- 1767 Published *History of Electricity*, with help and encouragement from Benjamin Franklin, whom he met in London.
- The same year became minister of a chapel in Leeds and started studying chemistry.
- 1772 Elected to the French Academy of Sciences; then to St. Petersburg Academy, 1780.
- Pioneer in the chemistry of gases.
- Discovered oxygen at the same time as Carl Scheele.
- 1774 Traveled to Europe accompanying Lord Shelburne and published *Letters to a Philosophical Unbeliever.*
- After moving to Birmingham then London, he moved to the United States in 1794.

Even before classical physics began to take shape and explore the forces that operated in the world of nature, the science of matter had made considerable progress in another direction by concentrating on the composition of matter itself. This science—chemistry—attempted to answer two ancient questions: Were all the innumerable different forms of matter really composed of just a few basic substances; and if so, how did they combine and separate? Chemists and alchemists before them had been accumulating isolated facts for centuries, but what they could not find was a theory of the processes involved—unless they were magical ones. The difficulties were enormous—measuring and testing substances that were always mixed together with other substances, trying to decide which was primary and which was secondary, and evolving a language of fundamental concepts that we now take for granted: element, compound, reaction, gas, oxidation, and so on.

THE HEAT IS ON

Combustion had long been seen as the main agent of chemical change, and the pioneers of the late eighteenth century tried to rationalize—give a rational explanation to—its effects step by step. As early as the 1750s Joseph Black (who also carried out important work in early heat theory, see above, page 6) succeeded in isolating the first scientifically recognized gas: carbon dioxide. Black did so by heating chalk and white magnesium, and collecting the gas that was given off. He called it simply "fixed air" because it seemed that it had been fixed inside those substances. He was intrigued to find that the process was reversible: that if the ashes of the chalk or magnesium were exposed to normal air, they reabsorbed the gas. This suggested to him that normal air contained a number of different gases, one of which was this fixed air. Black used very precise balances to show what was lost or gained during combustion, and he found that his "fixed air" was also produced by natural processes such as respiration and fermentation. This isolation of carbon dioxide, and the recognition of its role in a number of processes, can be said to mark the beginning of modern chemistry. Black acknowledged that chemistry was in its infancy, but emphasized that "experiment is the thread which will lead us out of the labyrinth."

The sequence of discovery continued with other gases. Hydrogen was isolated by the Englishman Henry Cavendish (1731–1810), who, like Black, also carried out important research in physics. Cavendish found that a highly inflammable gas was given off by various metals when they were attacked by hydrochloric acid. Was this gas perhaps phlogiston? It seemed unlikely since its parent substances were not themselves readily combustible. Cavendish called this gas "flammable air," and most important of all, he found that when it was burned—or rather exploded—with oxygen, a deposit of water was formed. Since both gases had been pure and unmixed, so far as he could ensure, Cavendish concluded that water was composed of hydrogen and oxygen, and that when they were mixed in the proportion 2:1, they formed their own weight of water.

The gas that forms the largest constituent in air, nitrogen, was isolated by the Edinburgh scientist Daniel Rutherford (1749–1819), who placed mice in a closed container until they absorbed all the oxygen, then exposed the remaining air to caustic potash, which took up the carbon dioxide. What remained was large by volume, but had few positive qualities, for it did not support respiration or combustion. Rutherford called it "mephitic air" from the Latin meaning "poisonous."

OXYGEN BREAKTHROUGH

The last, and perhaps most important, breakthrough in the identification of the gases was the separation of oxygen by an

Above: Joseph Priestley's house and laboratory in Birmingham where he carried out his historic experiments to isolate oxygen. This engraving by William Ellis is dated May 1, 1792. The house was attacked and burned by a mob incensed at Priestley's sympathy for the French Revolution.

Opposite: Henry Cavendish, discoverer of hydrogen.

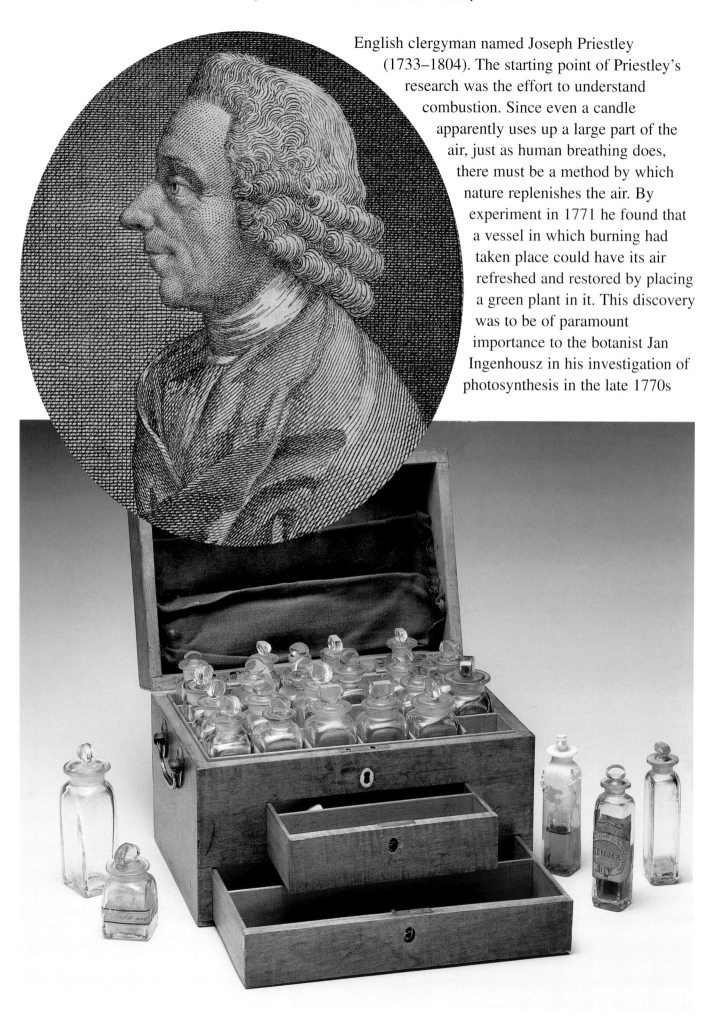

English clergyman named Joseph Priestley (1733–1804). The starting point of Priestley's research was the effort to understand combustion. Since even a candle apparently uses up a large part of the air, just as human breathing does, there must be a method by which nature replenishes the air. By experiment in 1771 he found that a vessel in which burning had taken place could have its air refreshed and restored by placing a green plant in it. This discovery was to be of paramount importance to the botanist Jan Ingenhousz in his investigation of photosynthesis in the late 1770s

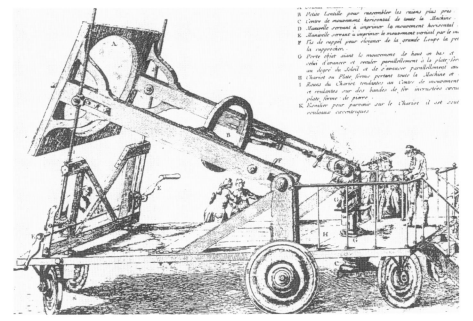

Left: The great lens constructed by the Paris Academy of Science; it was used to focus the Sun's rays to create intense but localized heat for chemical experiments.

Opposite, Above: Joseph Priestley, English-American theologian and chemist in a February 1, 1782, portrait.

Opposite, Below: A chemical chest of the period, this one owned by Michael Faraday.

(see Volume 6, page 15). Priestley experimented with heating various substances and collecting the gases that they gave off, until in 1774, on trying mercuric oxide, he discovered a gas that burned with a bright flame. He found that this was the gas that restored the air after combustion and respiration, indeed that mice and other small animals behaved more energetically than normal when exposed to it. Priestley was a firm believer in the phlogiston theory (see Volume 6, page 26), and he therefore called this gas "dephlogisticated air."

There is some evidence that other chemists were working along the same lines as Priestley; for example, in Sweden Carl Wilhelm Scheele reported his independent discovery of "fire air"—oxygen—in 1777.

Priestley continued his work, identifying a number of other gases, including ammonia and sulfur dioxide. He compared the properties of these gases, for example, by filling balloons with them and weighing them to determine their relative densities. He also invented soda water by passing carbon dioxide through water and noting the pleasant effervescent result. He suggested that it should be used on long sea voyages to make stale water more palatable.

This first generation of scientific chemists had succeeded in separating common air into some of its basic constituent gases. They had shown that these gases had very individual properties, and that they took part in physical and biological processes each in their very different ways. Yet they had not achieved a real conceptual breakthrough: They had not evolved a new language that could express systematically what was happening in their experiments, nor had they related the multitude of natural substances to a few basic components. That achievement was near, and it came through two of the most original thinkers in the history of chemistry, Lavoisier and Dalton.

Below: Carl Wilhelm Scheele, the Swedish chemist who was working in the same areas as Priestley.

The Theory of Elements:
Antoine Lavoisier

Antoine Lavoisier (1743–94) drew together the experimental work of scientists such as Black, Cavendish, and Priestley, and went on to place their results within a conceptual framework that enabled chemistry to move forward on a totally new level. Lavoisier was a wealthy man who devoted himself to scientific research. He worked for the French government on a number of projects, but during the French Revolution he was executed because of certain associations with the royal regime.

Lavoisier knew of Priestley's work, indeed, Priestley traveled to Paris in 1774–75 and exchanged ideas with the great French

A model of Lavoisier's laboratory in Paris around 1789.

scientist. Lavoisier repeated Priestley's experiments to obtain oxygen and then burned various substances in its presence and in common air. Meticulous measurements before and after the burning led Lavoisier to the revolutionary insight that what was happening in combustion was that oxygen was combining with the substances being burned, and that the phlogiston theory was a myth. He noted that sulfur and phosphorus, for example, when burned in air, increased in weight, and that part of the air was used up in the process. What remained was Black's "fixed air" and a large quantity of the inert gas that Rutherford had called "mephitic air." Lavoisier named the gas that he believed was essential to combustion "oxygen," meaning in Greek "acid-producing," from the mistaken belief that all acids contained oxygen. He named the largest constituent of air (Rutherford's nitrogen) "azote," again from the Greek meaning "no life." He agreed with Cavendish that water must be composed of hydrogen and oxygen. Hydrogen means "water-creator" and again follows Lavoisier's belief that chemical names should embody their origin or their chief function.

FUNDAMENTAL PRINCIPLES

But Lavoisier's genius went beyond explaining individual experiments. He concluded that all matter was composed of a small number of pure, elementary substances that could not be further subdivided, but which combined in an almost infinite variety of compounds. The classical idea of the four elements—earth, air, fire, and water—was finally laid to rest when Lavoisier published his *Traite elementaire de chimie* ("Treatise on the Elements of Chemistry") in 1789, for here he enumerated 23 of the recognized elements. Intriguingly, Lavoisier still listed "caloric," the imagined heat-fluid, as an element, and he regarded light also as an element.

He set out the two fundamental principles of modern chemistry: first that elements react with each other to produce compounds; and second that in these reactions matter is conserved. "In all the operations of art and nature," he wrote, "nothing is created: The quality and quantity of the elements remain the same, and nothing takes place beyond changes in the combinations of those elements." So convincing was Lavoisier's approach that within a few years the phlogiston theory had withered and died, and the task of the new chemistry was seen to be the isolation of

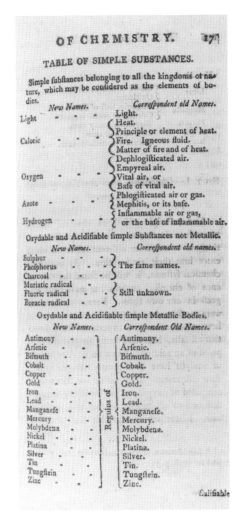

Above: Lavoisier's table of elements: Notice that he still believed light and caloric (heat) to be elements, alongside oxygen and hydrogen.

Below: Some of Lavoisier's equipment with which he analyzed chemical substances until they could be broken down no further—they were elements.

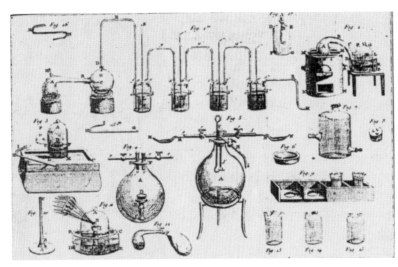

ANTOINE LAVOISIER (1743–94)

- Founder of modern chemistry.
- Born in Paris, France.
- 1768 Appointed as farmer-general of taxes; used his income for his researches.
- 1776 Became director of the French government's powder mills. He made gunpowder safer to produce by improving its quality as well as its supply and manufacture.
- Investigated the application of chemicals to improve agriculture.
- Helped reform the French tax system; worked to improve French prisons and hospitals.
- 1788 Using Joseph Priestley's investigations, he discovered oxygen, showed that air is a mixture of oxygen and nitrogen, realized oxygen's importance in breathing, its combustion potential, and its use as a compound with metals.
- 1789 Published *Traité élémentaire de chimie* ("Elementary Treatise on Chemistry"), a widely acclaimed discussion and explanation of chemical knowledge that founded chemistry as a modern science.
- Devised the modern way of naming chemical compounds.
- Was a member of the committee that devised the metric system.
- 1794 Guillotined on trumped-up charges of counterrevolutionary activities, but actually because he was hated as the chief collector of taxes.

the elements and the systematic analysis of their reactions with each other.

Although his life and career were tragically cut short, Lavoisier left papers which show that he was working on the application of chemistry to wider fields of nature. He was convinced that respiration was also a form of oxidation, and that oxygen was therefore deeply involved in the processes of life. He was also considering the natural cycles by which elements such as oxygen, nitrogen, and carbon are used and replaced in animal and plant life. In other words, he sensed that many apparently physical processes were essentially chemical in nature, and he was truly the father of modern chemistry.

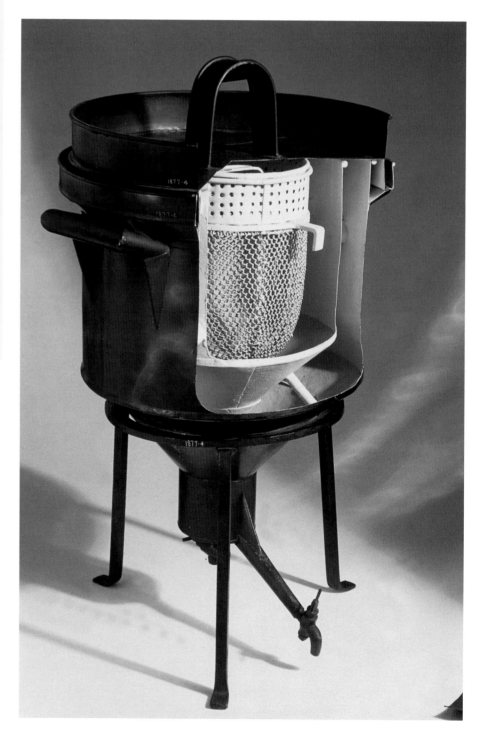

Right: Lavoisier's ice calorimeter. A guinea pig was placed in the central chamber, and the heat from its body melted ice placed around it. Lavoisier concluded that the heat in a living body came from oxidation, i.e., burning.

Opposite: Lavoisier and his wife, Marie-Anne Pierrette Paulze, 1788. She was his principal assistant. From a painting by Jacques-Louis David.

The Atomic Theory:
John Dalton

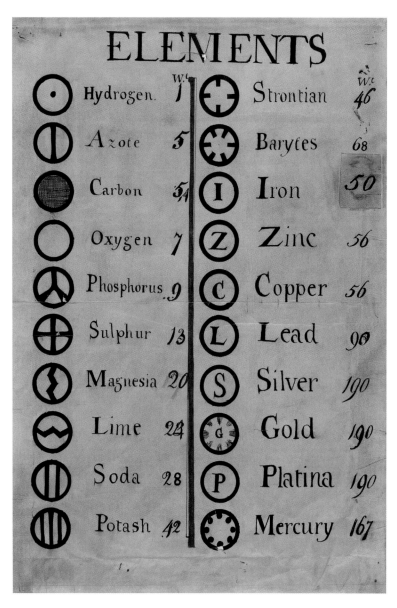

Above: Dalton's table of elements, with the graphic symbols that he devised, but that were not generally accepted.

Opposite: John Dalton, father of modern atomic chemical theory, from an etching by J. Stephenson in the 1820s. Dalton, among other things, first described color blindness—also known as Daltonism—an affliction from which he and his brother suffered.

What Lavoisier was unable to spell out was how and why certain elements combined to produce compounds while other do not, and why simply exposing two elements to each other does not necessarily produce a reaction. The most important single step on the road to answering these questions came when the English Quaker and teacher John Dalton (1766–1844) published his *New System of Chemical Philosophy* in 1808, in which the author revived the ancient idea of atoms and argued strongly that chemical reactions take place at the atomic level. The atom was simply a concept. No one, including Dalton, had any idea whether they really existed or what they were like, but they gave a logical explanation to chemical processes.

Interest in atoms as an idea had begun to revive in the late seventeenth century, partly from the discovery that air had weight and properties. That could be explained by supposing that it was composed of minute invisible particles so tiny as to be undetectable, yet adding up to a very definite physical mass. Robert Boyle held this view and extended it to liquids and solids. Newton stated the hypothetical existence of atoms very influentially when he wrote, "God in the beginning formed matter in solid, massy, hard, impenetrable particles . . . even so very hard as never to wear of break in pieces . . . ."

FIXED PROPORTIONS

A possible link between atoms and chemical combination had been suggested by a French contemporary of Lavoisier, Joseph-Louis Proust (1754–1826), in 1794, when he formulated his law of definite chemical proportions. Proust found that when he analyzed compounds into the parts that made them up, they always contained the same proportion by weight of their elements. For example,

Drawn & Etch'd by J. Stephenson.

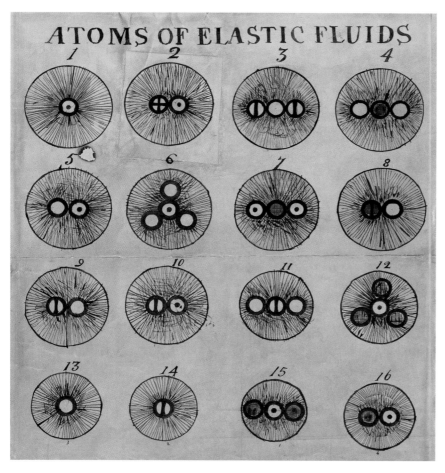

ATOMS OF ELASTIC FLUIDS

Two views of Dalton's scheme of atoms and molecules—1 is hydrogen, 2 is hydrogen sulphide, 3 is nitrous oxide, and so on; Dalton did not yet understand how complex molecular structure was.

Ether

Alcohol

copper carbonate was always five parts copper, four parts oxygen, and one part carbon. Dalton pondered these fixed mathematical proportions and at last concluded they were explicable if matter was combining at the atomic level. Minute particles of each element, having different weights, were combining one to one to produce a compound. The atom provided a unifying principle through which one could understand the constant proportions that always existed within a compound.

Dalton set out to estimate the weights of 20 elements, using hydrogen, the lightest, as base 1. This was the first attempt to define the element mathematically, for the weight of each element was unique. On this scale oxygen was 7, iron 50, and the heaviest was mercury at 167.

Dalton guessed that different compounds might be formed from the same elements if they combined in different proportions. For example, carbon dioxide contains twice as much oxygen as carbon monoxide. Unfortunately, Dalton rather confused the terminology by speaking of the combination itself as an atom, for example, an atom of water, although it was later more clearly described as a <u>molecule</u> composed of atoms. Dalton proposed several fundamental principles of atomic theory that have stood the test of time: that all matter is composed of atoms; that atoms of each element are identical to each other but different from those of other elements; that chemical reactions take place through the rearrangement of atoms; and that atoms are neither created nor destroyed.

Dalton proposed a system of graphic symbols to represent the elements and to show how they combined to produce their compounds. But they were rather clumsy and would have been difficult to follow in a complex equation. A different system was later developed. Despite certain misconceptions (for example, his idea of one-to-one combination, so that water in his system was HO), Dalton's theory was a work of real genius. It would later be proved by experimentation, and it would provide the foundation of the modern understanding of matter.

Forging a Chemical Language

The work of Lavoisier and Dalton represented two vital breakthroughs in creating a conceptual framework for chemistry, but a huge number of detailed questions still remained unanswered. In the years between 1810 and 1860 chemists throughout Europe tried to measure and quantify what was happening in chemical reactions in an effort to forge a consistent language of chemistry. The continuing uncertainty can be illustrated by the ideas put forward by the English chemist William Prout (1785–1850). Prout looked at the fact that hydrogen was the lightest element, and that the weights of all other elements were clear multiples of it, and on this basis he suggested that all the other elements might simply be compounds of hydrogen.

In France Joseph-Louis Gay-Lussac (1778–1850) made experiments that seemed to show that it was the volume proportions—the amount of each element—that were important in chemical compounds, not simply the atomic weights, as Dalton had believed. He found, for example, that when hydrogen and oxygen combined to form water, the proportions were two volumes of hydrogen to one of water, thus giving the formula for water as H_2O, rather than simply HO, as Dalton had stated.

LINKING UP

The reason for the disagreement between Dalton's approach and Gay-Lussac's was that the distinction between an atom and a molecule had not yet been understood. This step was taken by the Italian chemist Amedeo Avogadro (1776–1856), who linked the atomic theory with the volumetric approach by proposing that the atoms in certain elements, and those in all compounds, exist joined in an atomic group, which he named the "molecule" (Latin = "little mass"). Thus free oxygen became O, and only half a volume of oxygen was needed to combine with one volume of carbon monoxide to form one volume of carbon dioxide. It was molecules that were separated and reformed into new patterns during chemical reactions, while the atoms remained unchanged, and this was the true basis of what had been perceived as the conservation of matter. This was a crucial idea, but it was not widely accepted because these early chemists could calculate in only a very few cases how many atoms were in each molecule.

Above: Joseph Gay-Lussac.

Below: Electrochemical atoms, drawn by Berzelius in 1818. This diagram was the first attempt to explain electrolysis: When opposite charges faced one another (above), atoms combined; when similar charges faced one another (below), no reaction occurred.

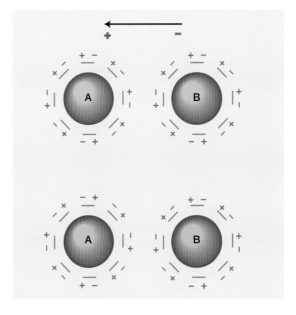

ILLUSTRATIONS OF NATURAL PHILOSOPHY.

CHEMISTRY.

TABLE OF THE ELEMENTS OF MATTER.

NON METALLIC ELEMENTS.

GAZOLITES.
(bodies permanently gaseous.)

Oxygen, Nytrogen, Hydrogen.

HALOGENS.
(bodies which produce Salts when in union with the metals.)

Chlorine, Iodine, Bromine, Fluorine.

METALLOIDS.
(bodies resembling metals in their chemical relations.)

Carbon, Boron, Silicon, Sulphur, Selenium, Phosphorus.

Oxygen, Chlorine, Bromine, Iodine and Fluorine having a tendency to combine with almost all other substances, and their union being generally accompanied by light and heat, have hence been termed Supporters of Combustion.

METALLIC ELEMENTS.

I

BASES OF THE EARTHS.
Alluminum, Thorium, Glucinium, Zirconium, Ittrium.

BASES OF THE ALKALINE EARTHS.
Calcium, Barium, Strontium, Magnesium.

BASES OF THE ALKALIES.
Potassium, Sodium, Lithium.

II

METALS WHICH DECOMPOSE WATER AT A RED HEAT.
Manganese, Zinc, Iron, Tin, Cadmium, Cobalt, Nickel.

METALS WHICH DO NOT DECOMPOSE WATER BY HEAT ALONE.
Arsenic, Antimony, Tellurium, Chromium, Vanadium, Uranium, Molybdenium, Tungsten, Columbrium, Titanium, Cerium, Bismuth, Lead, Didymium, Lantanium, Copper.

METALS WHOSE OXIDES ARE REDUCED BY A RED HEAT.
Gold, Silver, Mercury, Palladium, Rhodium, Platinum, Osmium, Iridium.

The above 55 Substances comprise the whole of the elements of matter, at present known.

Chemists have arranged the several forms of matter into the four following classes.

SOLIDS.
Which form the principal parts of the Globe, and differ from each other in hardness colour, density, &c.

FLUIDS.
As Water, Oil, &c. whose parts possess freedom of motion, and are generally non-elastic.

GASES.
Whose parts are highly moveable, elastic, transparent, varying in colour and greatly in density.

ETHEREAL SUBSTANCES
Known to us only by their motion, when acting upon our organs of sense, and which are not susceptible of being confined. Such are the rays of light, and radient heat.

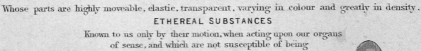
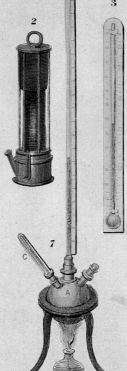
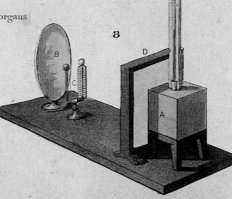

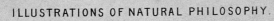

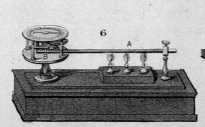

London Published by James Reynolds, 174 Strand, Dec 10, 1850.

Drawn & Engraved By John Emslie.

MAKING IT BALANCE

A great step forward came when the Swedish chemist Jons Berzelius (1779–1848) proposed using alphabetical symbols for the elements, usually based on their Latin names: Fe for iron, Au for gold, H for hydrogen, Pb for lead, and so on. These symbols proved universally acceptable where Dalton's graphic symbols had not, and they made possible the expression of chemical reactions as equations. In a chemical equation, as in a mathematical one, both sides must balance, and this balance expresses the idea of the conservation of matter through all its transformations.

By the midcentury the number of substances believed to be elements incapable of subdivision stood at around 50. The real existence of atoms was still not accepted by all scientists. Many considered that they were a convention, a working hypothesis only, not a physical reality. If chemistry was to develop—if it was to be taught in colleges, for example—it was necessary that its language should reflect a consistent understanding of its processes that all scientists could agree with.

In 1860 the first international chemical conference took place in Karlsruhe in Germany; at it Avogadro's ideas of atoms, composite atoms, and molecules were adopted, as well as the idea of volumetric analysis. Armed with this terminology and with the symbols proposed by Berzelius, the chemist could rationalize the results of his or her experiments, explaining how elements had joined to form new compounds and what byproducts might arise, all in a strictly agreed language.

Above: Sir Edward Frankland, who introduced the concept now known as valency.

Opposite: Chemical equipment seen in an engraving of December 10, 1850.

Below: The discovery of a new element, thallium, illustrated by six specimens. Thallium was discovered in 1861 by William Crookes (1832–1919), who isolated it by its green line. Thallium is a rare soft white metallic element, occurring naturally in zinc blende and some iron ores.

DE L'EXISTENCE D'UN NOUVEAU METAL,

LE THALLIUM

Par M. A. LAMY,

Above: Friedrich August Kekulé, German organic chemist and discoverer of benzene rings.

Right: William Hyde Wollaston's (1766–1828) models of crystals from the early nineteenth century. These wooden models show how groups of spherical atoms can form crystal shapes

Below: Robert Bunsen, inventor of the burner named for him, with his colleagues Gustav Kirchhoff and Henry Roscoe.

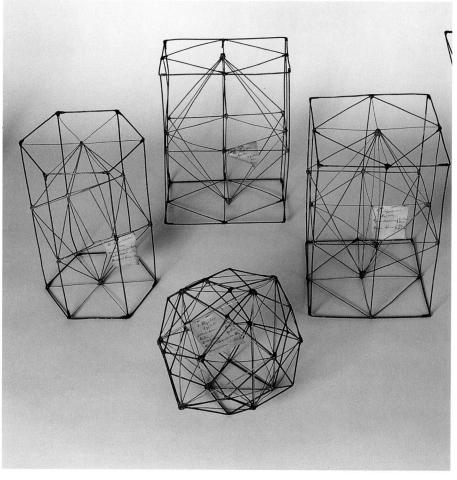

CHEMICAL BONDS

Another issue of discussion at Karlsruhe was the concept of valency. The English chemist Edward Frankland (1825–99) first deduced this property, arguing that the atoms of each element will always combine with a certain number of other atoms: hydrogen with one, oxygen with two, nitrogen with three, carbon with four, and so on. The German Friedrich Kekulé (1829–96) realized that it was a peculiar property of carbon that one of its chemical bonds could be used to combine with another carbon atom, and that in this way a long, complex molecule could result. This discovery became of great importance in organic chemistry, which studies the properties of these large carbon molecules.

The results of the Karlsruhe conference were the culmination of the revolution in chemical ideas that had begun with Lavoisier in the 1770s. The question of the physical nature of atoms and the forces that held them together still remained as baffling as ever, but they came to be seen as the province of physics. Chemistry would see its role as systematizing the thousands and thousands of possible transformations that the elements might undergo. The chemical bond was understood to be electrical in character, as shown by the experiments of Volta, Davy, Faraday, and others, and again it became the province of physics to investigate the mysterious qualities of that bond.

COLOR CODING

One of the most important experimental advances came in 1859 with the advent of <u>spectroscopy</u>. Two German scientists, Robert Bunsen (1811–99) and Gustav Kirchhoff (1824–87), discovered that when different substances burn, their light could be analyzed with a prism to show their different colors and characteristics, which were as distinctive as a system of fingerprints. The results of spectroscopy could be recorded and compared, providing a powerful analytical tool with which substances could be identified. As the science of matter's inner composition, chemistry had proved the most complex challenge for scientists to overcome, but by 1860 its foundations were secure, and its lessons were being applied in quite dramatic ways to create new industries and new materials.

ROBERT BUNSEN (1811–99)
- Chemist and physicist.
- Born in Göttingen, Germany.
- Went to school and college in Göttingen, then on to Paris, Berlin, and Vienna.
- Appointed professor at Kassel, then Marburg, Breslau, until finally settled as a professor at Heidelberg.
- Adapted the "Bunsen" burner from an invention of Michael Faraday's and experimented widely using it.
- Invented the grease-spot photometer, an ice calorimeter, a galvanic battery, and—in colaboration with Henry Roscoe—an actinometer.
- Made a comprehensive study of organoarsenic compounds, gas analysis, and electrolysis.
- 1859 Discovered (with Gustav Kirchhoff) spectrum analysis. This helped in the discovery of new elements.
- In a laboratory accident during an experiment partially lost the use of one eye—from then on forbade the study of organic chemistry in his lab.

Left: A selection of Bunsen burners from the second half of the nineteenth century. Bunsen invented the burner in 1855.

Below: Baron Jons Jacob Berzelius, the Swedish chemist who discovered the elements selenium, thorium, and cerium.

The Periodic Table of the Elements: Dmitri Mendeleev

Above: Dmitri Mendeleev, who discovered the fundamental order that exists among the chemical elements.

Opposite: One of Mendeleev's original sketches for his periodic table, setting out the elements in related groups.

In trying to probe the mystery of the atom, scientists felt that the one certain fact that could be stated about it was its atomic weight. Of course, atoms are so small that it was not (and still is not) possible to weigh a single atom, but the term really meant the relative mass of different substances—carbon 12 times that of hydrogen, for example. Such relative masses were possible to calculate, and it was Jons Berzelius who determined the atomic weights of many elements. It was perhaps strange that it was not until the 1860s that a chemist began to study these weights to see if they revealed anything significant about the element.

The chemist in question was the Russian Dmitri Mendeleev (1843–1907), who had studied with the leading chemists in France and Germany, and had attended the Karlsruhe conference before being appointed professor at St. Petersburg University. Mendeleev could find no textbook suitable for introducing students to the principles of chemistry, so he set out to write his own. The task of organizing a book and of rationalizing chemical theory made him search for ideas and principles that would bring order to the complex world of chemicals. He felt strongly that the atomic weights of the different elements could not be accidental, and he began to set out tables of the elements in ascending order of their weights and looking at their relationships.

ALL IN ORDER

His instinct was correct, for he saw at once that the elements fell into groups that shared important characteristics. For example, the heavy metals, gold, mercury, and lead, appear in sequence in the table, as do the alkali metals, which do not occur freely in nature, such as calcium and potassium. Another clearly defined group consists of the light, corrosion-free metals titanium, vanadium, and chromium.

Mendeleev drew the conclusion that these groups of elements shared characteristics because their atomic weights were similar,

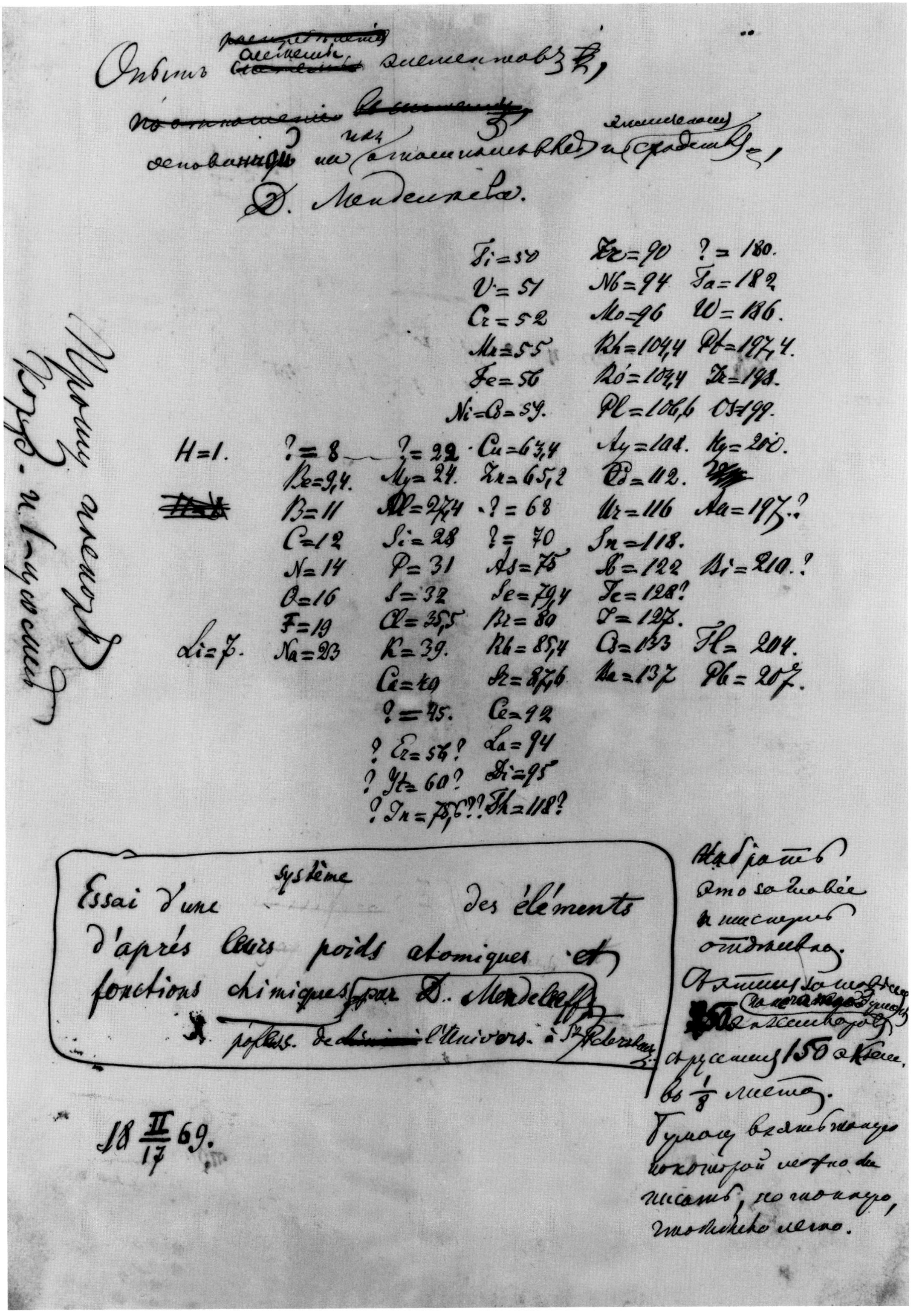

Опытъ системы элементовъ.

основанной на ихъ атомномъ вѣсѣ и химическомъ сходствѣ.

Д. Менделѣева.

				Ti = 50	Zr = 90
				V = 51	Nb = 94
				Cr = 52	Mo = 96

$Ti = 50 \quad Zr = 90 \quad ? = 180.$
$V = 51 \quad Nb = 94 \quad Ta = 182.$
$Cr = 52 \quad Mo = 96 \quad W = 186.$
$Mn = 55 \quad Rh = 104,4 \quad Pt = 197,4.$
$Fe = 56 \quad Ro = 104,4 \quad Ir = 198.$
$Ni = Co = 59. \quad Pl = 106,6 \quad Os = 199.$

$H = 1.$

$? = 8 \quad ? = 22 \quad Cu = 63,4 \quad Ag = 108. \quad Hg = 200.$
$Be = 9,4 \quad Mg = 24. \quad Zn = 65,2 \quad Cd = 112.$
$B = 11 \quad Al = 27,4 \quad ? = 68 \quad Ur = 116 \quad Au = 197 ?$
$C = 12 \quad Si = 28 \quad ? = 70 \quad Sn = 118.$
$N = 14 \quad P = 31 \quad As = 75 \quad Sb = 122 \quad Bi = 210. ?$
$O = 16 \quad S = 32 \quad Se = 79,4 \quad Te = 128?$
$F = 19 \quad Cl = 35,5 \quad Br = 80 \quad I = 127.$
$Li = 7. \quad Na = 23 \quad K = 39. \quad Rb = 85,4 \quad Cs = 133 \quad Tl = 204.$
$Ca = 40 \quad Sr = 87,6 \quad Ba = 137 \quad Pb = 207.$
$? = 45. \quad Ce = 92$
$? Er = 56 ? \quad La = 94$
$? Yt = 60? \quad Di = 95$
$? In = 75,6 ?? \quad Th = 118?$

Essai d'une système des éléments
d'après leurs poids atomiques et
fonctions chimiques par D. Mendeleeff
роков. de l'Univers. à St Petersbourg.

Набрать это заглавіе и напечатать отдѣльно.

$18 \frac{II}{17} 69.$

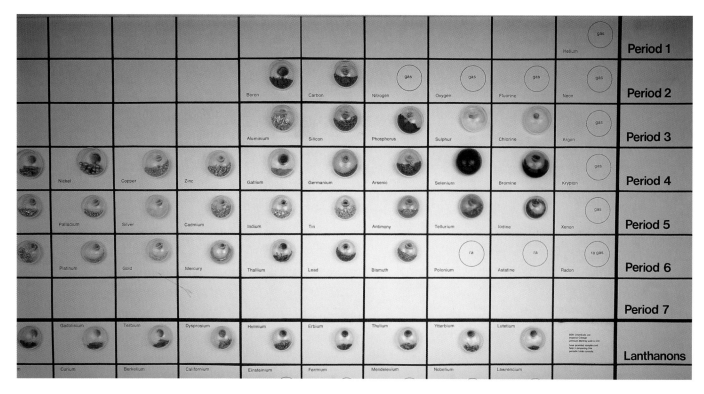

													Period 1
				Boron	Carbon	Nitrogen (gas)	Oxygen (gas)	Fluorine (gas)	Neon (gas)				Period 2
				Aluminium	Silicon	Phosphorus	Sulphur	Chlorine	Argon (gas)				Period 3
Nickel	Copper	Zinc		Gallium	Germanium	Arsenic	Selenium	Bromine	Krypton (gas)				Period 4
Palladium	Silver	Cadmium		Indium	Tin	Antimony	Tellurium	Iodine	Xenon (gas)				Period 5
Platinum	Gold	Mercury		Thallium	Lead	Bismuth	Polonium (ra)	Astatine (ra)	Radon (ra gas)				Period 6
													Period 7
Gadolinium	Terbium	Dysprosium	Holmium	Erbium	Thulium	Ytterbium	Lutetium						Lanthanons
Curium	Berkelium	Californium	Einsteinium	Fermium	Mendelevium	Nobelium	Lawrencium						

Sample elements arranged in periodic table formation.

Mendeleev in old age.

and that as a sequence was traced, there would be a gradual transition in chemical properties. He called his list a "periodic" table, meaning numerical, mathematical, regular: The atomic number of the element did indeed signify something objective about the substance. This was almost like a revival of the ancient number mysticism of Pythagoras and Plato, who believed that mathematical order had been built into the structure of the universe (see Volume 1, pages 38–44).

Mendeleev's table became a fundamental organizing tool for chemists, bringing order into the bewildering variety of material substances. The table was incomplete, but he predicted that its gaps would be filled in time by the discovery of new elements, and in this he was proved correct when the metals gallium, germanium, and scandium were discovered in the 1870s and 1880s, and the inert gases helium, neon, and argon in the 1890s. Mendeleev had been occupied for years in the search for unifying principles in chemistry, and then he formulated the theory of his brilliant discovery in a single day, March 1, 1869.

We now know that the atomic number actually derives from the number of protons (or electrons) in the atom of each element, but this knowledge was not available to Mendeleev. In the atomic age it became possible to transform one element into another by the addition or deletion of an electron, in other words, moving the element up or down the periodic table. Mendeleev's achievement was perhaps the highpoint of the nineteenth-century investigation of the atom, which could not be seen or examined directly, but whose existence and nature could only be probed by reason and by drawing conclusions from careful experiments.

The Rise of Organic Chemistry: Justus Liebig

Justus Liebig (1803–73) was born into the world of chemistry, for his father was a dealer in dyes, drugs, and associated chemicals in the German city of Darmstadt. After studying in Paris under Gay-Lussac, the brilliant young Liebig was appointed professor of chemistry at the University of Giessen at the age of 21. Here he developed one of the first university teaching laboratories, and over the next 30 years he trained hundreds of young scientists in the language of chemistry and the techniques of analysis. Liebig is credited with developing the entire realm of <u>organic chemistry</u>, that is, the chemistry involving carbon and its compounds, which are fundamental to the processes of life. Although not widely distributed in nature, forming only about 0.2 percent of the Earth's crust, carbon is unique among the elements in its ability to form a huge number of compounds built of large, complex molecules, most often also involving oxygen and hydrogen. The chemists of the late eighteenth century found carbon in the vital fluids of the body, such as blood, milk, sweat,

Above: Justus Liebig, pioneer of organic chemistry.

Left: The carbon cycle links plant and animal life, and it was one of Liebig's outstanding discoveries.

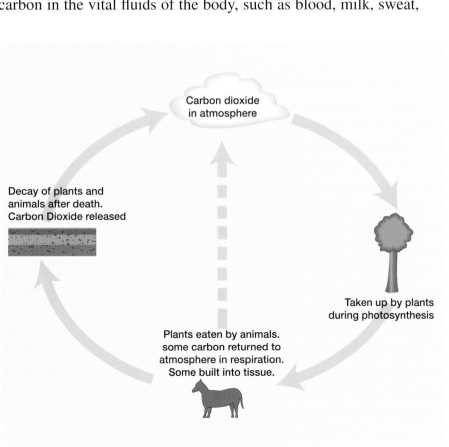

Carbon dioxide in atmosphere

Decay of plants and animals after death. Carbon Dioxide released

Taken up by plants during photosynthesis

Plants eaten by animals. some carbon returned to atmosphere in respiration. Some built into tissue.

Opposite: A mid-nineteenth-century Liebig condenser that was made by J. Newton in London, England. Liebig showed that the organic world was simply a more complex manifestation of the same chemicals that made up the inorganic world.

and urine, and they had also demonstrated that it was exhaled as carbon dioxide. But how did it get into the body?

OUT OF THIN AIR

To answer this question, Liebig turned his attention to agriculture and looked at the whole relationship between soil, plants, and animals. All agriculturalists had believed that plants absorbed carbon from humus, that is, from decomposing plant material. Liebig, however, carried out experiments that produced some very remarkable results that disproved this. He showed that a given area of land, whether it was cultivated field or forest, contributed the same amount of carbon each year to the composition of the plants grown there. He deduced from this that plants extracted their carbon from the atmosphere, not from the humus. The stability of the carbon in the atmosphere, despite the continuous exhalation of carbon dioxide by animals, proves that there is some natural mechanism for recycling it. Photosynthesis had been described and experimentally proved by Jan Ingenhousz (see Volume 6, page 15), but he had not been able to give it a precise chemical formulation. The fundamental process carried out in plants, Liebig now stated, was to separate the carbon and the oxygen in carbon dioxide, to release the oxygen into the atmosphere, and to assimilate the carbon into their own tissue in the form of sugar and proteins (although the investigation of proteins was then only in its infancy).

Liebig was able to synthesize in the laboratory such important organic compounds as urea, thus demonstrating that living processes utilized the normal chemical elements. He also showed that the carbon cycle was the fundamental process that maintains life on Earth. All living organisms convert food into energy and into

Below: Liebig's laboratory in Giessen, where he trained hundreds of young scientists and advanced pure and industrial chemistry more than any other individual.

structural tissue, and the basis of this food is carbon, absorbed by plants during photosynthesis and then eaten by animals. Without the carbon cycle the food supply would halt, and the Earth would become devoid of oxygen.

INNOVATION FOR AGRICULTURE

Liebig then turned his attention to nitrogen. Nitrogen constitutes around 80 percent of the Earth's atmosphere and is another basic component of living tissue, yet it is notoriously unreactive. So how does it enter the bodies of plants and animals? Liebig's experiments led him to the discovery that nitrogen is contained in rainwater in the form of ammonia, which is highly soluble. In the soil it forms nitrates, which are then absorbed by plants and in turn eaten by animals. Liebig was an important pioneer of the idea of adding nitrate fertilizers to the soil. He was not able to give a complete account of the nitrogen cycle because he did not know of the part played by bacteria in returning the nitrogen to the air.

Liebig's work, however, was the beginning of a revolution in the theory of agriculture. Before 1840 it was believed that plant and animal life depended on the circulation of organic, previously living material. Liebig showed that the constituents and the nutrients of plants and animals were the normal inorganic elements, carbon, nitrogen, and so on. In the old view the potential for the growth of foodstuffs clearly had a fixed limit, but in Liebig's opinion the addition of basic inorganic elements to the soil offered an almost limitless scope for increased food production.

Liebig was also among a group of nineteenth-century scientists (including Helmholtz, see above, page 9) who were able to demonstrate that animal work and metabolism—the production of energy and tissue growth—were fueled by oxidation, in other words, by a standard chemical reaction among elements. This application of chemistry to physiology was, in a sense, the final defeat of any vitalist philosophy, of the idea that life was a mysterious force above and beyond any physical basis. It represented the first beginnings of biochemistry and also of scientific ecology—of the realization that plant and animal life and the environment of the Earth are all linked in one interdependent system.

JUSTUS VON LIEBIG (1803–73)
- Chemist.
- Born in Darmstadt, Germany.
- Studied in Bonn, then Erlangen.
- 1822 Moved to Paris and studied analysis techniques with Joseph Gay-Lussac.
- 1824 Appointed extraordinary professor at Giessen University. Set up an institute for training chemists.
- Studied the phenomenon of isomerism—substances that have the same molecular formula but with atoms differently connected.
- Developed improved procedures for elemental analysis of organic compounds; investigated organic, animal, and agricultural chemistry.
- Liebig's condenser—distillation equipment for carrying out chemical analysis.
- 1840 Published *Die organische Chemie in Ihre Anwendung auf Agricultur und Physiologie* ("Organic Chemistry and its Use in Agriculture and Physiography"). This led to great improvements in agriculture.
- 1852 Moved to Munich as professor of chemistry.

Astronomy:
The Changing Solar System

The main achievement of nineteenth-century astronomy was in its investigation of the universe of the stars and in revising people's ideas about the scale of the universe. But the more traditional study of the solar system could still produce its own revolutionary discoveries.

In 1772 two German astronomers, Johann Titius (1729–96) and Johann Bode (1747–1826), had drawn attention to a curious sequence in the planets' distances from the Sun. The numerical sequence runs:

$$0 + 4 = 4, 3 + 4 = 7, 6 + 4 = 10, 12 + 4 = 16,$$
$$24 + 4 = 28, 48 + 4 = 52, 96 + 4 = 100, 192 + 4 = 196.$$

If the totals in each of these sums are divided by 10, the result is extremely close to the distance of each planet from the Sun measured in astronomical units, that is, the Earth's distance from the Sun:

Mercury: 0.4, Venus: 0.7, Earth: 1.0, Mars 1.6,
Jupiter: 5.2, Saturn: 9.5, Uranus: 19.2.

John Couch Adams (1819–92) was a British astronomer who, in 1845, deduced mathematically the existence and location of the planet Neptune. Adams was appointed professor of astronomy at Cambridge in 1858 and was director of the Cambridge Observatory from 1861. The image is taken from a collection of his papers in 1896.

The fit is generally very good, but there was evidently a gap between Mars and Jupiter, and it was naturally asked whether an undiscovered planet could possibly be orbiting the Sun at 2.8 astronomical units distance. For some years astronomers studied this possibility, and in 1801 the Italian observer Giuseppe Piazzi (1746–1826) did indeed locate a planetary body in exactly the predicted orbit; he named it Ceres. Yet it proved to be very small, having a diameter of only 600 miles—roughly the size of Spain. The situation became still more puzzling when, in the years that followed, other even smaller bodies were found to be sharing the same orbit. By 1872 more than 100 "asteroids" had been located, and it was long believed that they must form the remains of a large planet that had somehow disintegrated. The status of the Bode-Titius law is still unclear. Neptune does not fit the pattern well, and with Pluto it breaks down completely. It appears to be a purely numerical relationship and not a physical law at all, but it had played a part in the discovery of a previously unknown feature of the solar system.

The "Alarming Comet of 1835" published in 1839, a lithograph by A. Ducote. It shows the comet—Halley's comet—with the face of Sir Edmund Halley, who calculated the comet's orbit. He had determined that the comets of 1531 and 1607 were the same object with a 76-year orbit. Halley didn't live to see his prediction come true, dying in 1742, 16 years before the next sighting of his comet.

NEW PLANET

In the 1820s astronomers also began to be puzzled by the orbit of Uranus, the outermost of the known planets, discovered by William

URBAIN LEVERRIER (1811–77)

- Astronomer.
- Born in St. Lô, France.
- 1836 Became a teacher of astronomy at the Polytechnique.
- Interested in the motions of the planets, he published *Tables de Mercure* ("Tables of Mercury") about the planet Mercury.
- 1846 Admitted to the French Academy.
- Deduced the existence of an unseen planet by observing irregularities in the motion of the other planets, he calculated its exact position; a few days later, exactly where he predicted, Johann Galle discovered the planet Neptune, 1846.
- 1849 Elected to the French Legislative Assembly.
- 1852 Napoleon III made him a senator.
- 1854–70 Worked as director of the Paris Observatory.

Herschel in 1781. Its orbit displayed small irregularities that suggested either that Newton's laws of gravitation did not apply rigidly at such a great distance from the Sun, or that the orbit was being disturbed by another body. Could there be another planet beyond Uranus? Using the mathematical models set out by Laplace (see Volume 6, page 11), astronomers set to work to predict the likely position of such a body and then to scan the night sky in the hopes of finding it. This led to one of the most contested "first" disputes in the history of astronomy. In September 1846 the French astronomer Urbain Leverrier (1811–77) submitted detailed predictions of the unknown planet's path to the Berlin Observatory to be used as a basis for searching, and within days the new planet was found.

But the international celebrations that followed became confused when it was announced that a young English mathematician, John Couth Adams (1819–92), had reached exactly the same result a whole year earlier. Adams had sent his calculations to the professor of astronomy at Cambridge University and to the astronomer royal, but they had not followed them up with an optical search of the sky. National rivalries between English and French scientists played a part in this story, but today both men are credited with the discovery of the new planet, which was named Neptune, god of the deeps of the sea, because it lay so deep in space.

WRITTEN IN THE STARS

There were two curious sequels to this discovery. First, Leverrier turned his attention to the orbit of Mercury, which had also shown

Right: The astronomical observatory at Cambridge University in 1825.

Opposite: Zodiacal light, dust from the tails of comets and from disintegrated asteroids, which sometimes appears along the ecliptic.

certain irregularities that had puzzled Newton and many other scientists. Leverrier predicted that another planet must lie inside Mercury's orbit, even closer to the Sun, and he decided to name it Vulcan after the god of fire; it was never found.

The second consequence was of little direct importance to science, but it affected the astrological community. Traditional astrological ideas about the influence of the planets had always been concentrated on the five planets known since ancient times. The discovery of Uranus and then of Neptune at first raised huge problems for astrologers: Surely these planets must also influence human affairs, and surely all previous astrology was made invalid because its practitioners had not even known the true number of planets.

Since the late nineteenth century astrologers have set about describing the characteristic influences of the new planets (and of Pluto), but those characteristics relate to the role of the gods Uranus, Neptune, and Pluto in classical mythology. However, we know that the naming of these new planets was really an arbitrary process: They might easily have been given different names and therefore would presumably have been given different qualities by the astrological community. This problem of the new planets was to prove a serious one for modern astrologers.

Probing the Universe of Stars

Opposite: Constellations of the northern celestial hemisphere of around 1850.

William Herschel had laid the foundations of a new approach to stellar astronomy (see Volume 6, page 56), and the leading astronomers of the nineteenth century devoted themselves to the task he had started—redrawing people's intellectual map of the universe. Herschel's son John (1792–1871) continued his father's work to the extent of observing and cataloging 2,300 nebulas by the year 1833. In the process he made a striking and important drawing of the nebula M51, which lies in the constellation Canes Venatici. He saw it as a central cluster surrounded by a divided ring of stars. He realized that this ring, if seen by an observer at the center, would be similar to the Milky Way as seen from the Earth. "Perhaps," he wrote, "this is our brother system."

The older Herschel had seen only the stars visible from England; but in order to complete his father's work, John transported his great 20-foot reflecting telescope in 1833 to a site near Cape Town in South Africa, where he spent the next four years surveying the southern skies. On his return to England he published a new catalog of 1,700 more nebulas, making it appear, as one commentator remarked, that the study of the nebulous heavens "was the exclusive domain of the Herschel family."

Above: Photograph of Sir John Frederick William Herschel, the English astronomer, taken in the 1860s. Sir John was the son of Sir William Herschel, another astronomer, and extended his father's work on the nebulas. He was also a pioneer photographer.

SEEING STARS

This private monopoly was broken in the early 1840s when William Parsons, later earl of Rosse, built a gigantic reflecting telescope at his home in Birr, Ireland, with a mirror six feet in diameter, easily the most powerful instrument of its time. Within weeks of its first use Rosse had produced a drawing of the M51 nebula superior to that by Herschel and showing its characteristic spiral form. The great question about these nebulas was whether they were really composed of stars, or whether they

CONSTELLATIONS

OF

THE NORTHERN

STARS OF THE 1st ★ 2nd ✴ 3rd ✦ 4th • MAGNITUDES

CELESTIAL HEMISPHERE.

LONDON: JAMES REYNOLDS & SONS, 174, STRAND.

Right: Rosse's original drawing of the M51 nebula, clearly showing its spiral form.

Opposite: Two views of Rosse's giant reflecting telescope. The mirror was 72 inches across, and Rosse made important observations with it, despite the cloudy and unfavorable climate of central Ireland where it was located.

were simply clouds of gas. Both Herschel and Rosse had clearly resolved M51 into a mass of stars, and its distinctive shape suggested strongly that it was a star system. But was it a subsystem within our own or separate from our own? There was still no way of answering these questions because no method of measuring or even estimating stellar distances had been discovered.

A significant step in that direction occurred when in 1838 the German astronomer Friedrich Wilhelm Bessel (1784–1864) succeeded in doing what had eluded all astronomers to that date, namely, measuring the parallax movement of a star caused by the Earth's motion in its orbit around the Sun. With the best instruments available Bessel carefully observed a star in the constellation Cygnus for more than a year before concluding that it showed a parallax movement of one-third of one second of arc. This distance then became the base of a very thin triangle, two of whose angles were known. When that base was stated as the diameter of the Earth's orbit, it became possible to derive a tentative figure for the star's distance from the Earth; Bessel gave it as just over 60,000,000,000,000 miles. In modern terminology this equates to around 12 light-years and is a very good estimate.

Within a year or two other stellar parallaxes were found. A Scottish astronomer, Thomas Henderson (1798–1844), then working at the Cape Observatory in South Africa, announced that the brightest star in the constellation Centaur had a parallax more than double that of Bessel's chosen star, meaning that its distance from the Earth was less than half, as indeed it is: The star Alpha Centauri, at four and a half light-years, is generally agreed to be the star closest to Earth. The magnitudes of these figures were beyond anything that scientists had anticipated, and they were for stars that were relatively close to the Earth. There seemed to be no answers in sight to the problem of comprehending the scale of the universe.

Unlocking the Chemistry of the Universe

Gustav Kirchhoff, one of the founders of stellar spectroscopy.

Fox Talbot, pioneer photographer who explained how burning chemicals could be analyzed prismatically.

In 1844 the French philosopher Auguste Comte (1798–1857) was discussing the limitations of scientific knowledge, and he gave his opinion that man would never know anything about the stars except what they looked like from a distance; we could never know about their physical or chemical natures. Within fewer than 20 years of this prediction Comte was to be proved wrong by one of the most important, and perhaps unplanned and unexpected, technical breakthroughs in astronomy—the technique of spectroscopy.

The fact that light from flames can be analyzed by a prism and will show different colors depending on the substance being burned was realized by a number of scientists early in the nineteenth century. In 1814 a Munich instrument maker, Joseph Fraunhofer, noticed that in these cases the spectrum was intersected by a number of dark lines whose meaning he was unable to explain, although he saw that the pattern of the lines varied as different substances were burned.

COLOR THEORY

The chemical significance of the spectrum was grasped by, among others, two Englishmen, the physicist and photographic pioneer William Fox Talbot (1800–77) and astronomer John Herschel (1792–1871). Talbot wrote that "a glance at the prismatic spectrum of a flame may show it to contain substances which it would otherwise require a laborious chemical analysis to detect."

In Germany Robert Bunsen (1811–99) developed his famous gas burner to burn with a colorless flame so that it would not interfere with this color analysis. It was his colleague Gustav Kirchhoff (1824–87) who suggested that he should refract the light through a prism instead of looking at it through various colored filters. These two chemists, working in Heidelberg in the 1850s with their Bunsen burner and prism, realized that each element as it burned emitted a spectrum of bright colors intersected by dark lines as individual as a fingerprint, and they discovered that this technique could be transferred from combustion in a laboratory to the realm of the stars.

The story is told that this discovery followed their observation of a distant fire that they saw from their laboratory window. With their spectroscope they were able to tell that among the substances that were burning were the metals barium and strontium. Some time later Bunsen suggested to Kirchhoff that if they could analyze the material in a distant fire, why should they not do the same for the

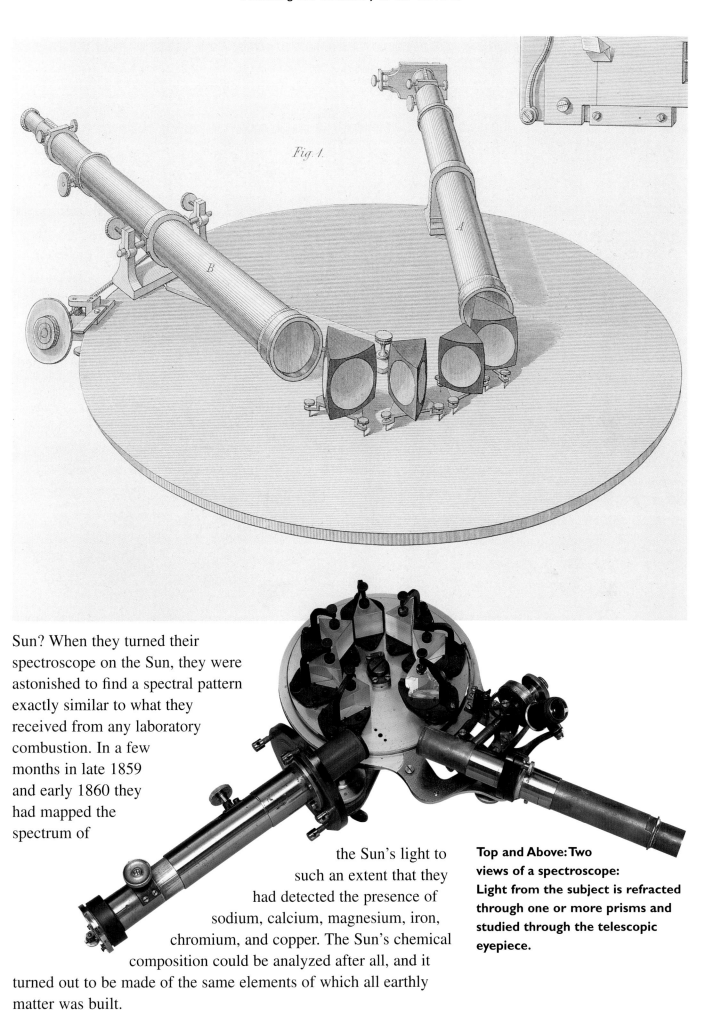

Fig. 1.

Sun? When they turned their spectroscope on the Sun, they were astonished to find a spectral pattern exactly similar to what they received from any laboratory combustion. In a few months in late 1859 and early 1860 they had mapped the spectrum of the Sun's light to such an extent that they had detected the presence of sodium, calcium, magnesium, iron, chromium, and copper. The Sun's chemical composition could be analyzed after all, and it turned out to be made of the same elements of which all earthly matter was built.

Top and Above: Two views of a spectroscope: Light from the subject is refracted through one or more prisms and studied through the telescopic eyepiece.

Christian Doppler, who discovered that a moving light source produces a spectral shift—a vital tool for stellar astronomers.

SPECTRAL ANALYSIS

A further startling conclusion was that the Sun (and therefore by extension the other stars) had an atmosphere. This discovery arose from the dark lines ("absorption lines") that crossed the colors of the spectra. Kirchhoff realized that these lines must be caused by bands of light of a given wavelength being absorbed by the same elements that existed as gases in the intense heat surrounding the Sun. When the light from burning sodium or magnesium, for example, encountered light from those elements burning as gases, the two canceled each other and produced the characteristic dark "absorption lines." It was the pattern of these lines combined with the pattern of colors in the spectrum that functioned like the fingerprint of any given star. The existence of the Sun's atmosphere came to be appreciated when the first photographs of the Sun revealed its huge glowing corona, which is its atmosphere, extending for millions of miles into space.

Kirchhoff wrote a series of papers in which he explained their discovery and rationalized what was being seen in the different spectra. A monument to Kirchhoff that stands in Heidelberg commemorates the discovery of spectral analysis that "unlocked the chemistry of the universe." Within a few years astronomer Norman Lockyer (1836–1920), in England had used the spectrum to identify an unknown element in the Sun, which he named helium, and which was not isolated on Earth until 1895. Spectroscopy was to yield another momentous discovery—that it could be used to measure the velocity with which the stars were moving.

DOPPLER EFFECT

This was an application of the principle described by the Austrian physicist Christian Doppler (1803–53) in 1842 that a moving source of energy produces radiation of a longer wavelength if its receding and of a shorter wavelength if it is approaching. That is because the waves of radiation do not arrange themselves symmetrically around a moving energy source: They are "squeezed" in front of it and "stretched out" behind it. When applied to the light from a star that is receding from the Earth, this means that the long-wavelength part, the red end of the spectrum, is stretched, while the short-wavelength blue end is contracted. This principle has become

A spectrum of sunlight: Different stars produce different patterns of dark lines ("absorption lines") depending on the chemicals present; these patterns act like a fingerprint showing the constitution of the stars.

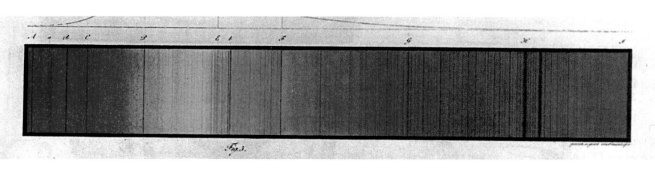

known to science as the "red shift." If a star were rushing toward the Earth at great speed, the reverse effect would occur—there would be a blue shift.

In 1868 the English astronomer William Huggins (1824–1910) analyzed the light of Sirius, the brightest star in the sky, and found that its red was shifted to the extent that he was able to calculate that the star was moving away from the Earth at a speed of almost 30 miles per second. This technique of analyzing the red shift was to assume the utmost importance in the twentieth century, when it was applied to galaxies outside our own. Spectroscopy became one of the basic tools of the astronomer when examining any star, and its findings helped build an entirely new and unexpected view of the cosmos.

TRAVELING LIGHT

Nineteenth century astronomy can be seen as an unfolding series of clues or stepping stones toward a new view of the cosmos, which finally emerged in the 1920s and 1930s. Another was the so-called "Olbers Paradox," named for the German astronomer Heinrich Olbers (1758–1840), although he was not the first to discuss it. Olbers asked simply why the sky is dark at night, given that the heavens are infinite, and the distribution of stars is even. The amount of light reaching the Earth from distant stars is very small, but on the other hand, the number of stars also increases with distance, so why isn't the night sky glowing with an even light?

Olbers thought there might be a kind of dust in space that absorbs the light, but the answer was divined by another German, Johann Mädler (1794–1874). The light from all the stars that were out there had simply not had time to reach the Earth. But if this were true, then the time of the light's travel must be less than the age of the universe, and given the immense speed of light, this had profound implications for ideas about both the age and the size of the universe.

Above: William Huggins, the British astronomer who measured the recession movement of Sirius.

Below Left: Olbers paradox. Stars are evenly scattered through space, and although the distant ones are fainter, there are more of them; therefore, the night sky should be evenly lit with starlight.

HEINRICH OLBERS (1758–1840)
- Astronomer.
- Born in Ardbergen, Germany.
- Studied medicine at Göttingen and Vienna.
- 1779 Worked out a way of calculating the orbits of comets.
- 1781 Set up in medical practice in Bremen, but studied astronomy at every opportunity from his small home observatory.
- 1802 Discovered the minor planet Pallas and another, Vesta, in 1807.
- 1815 Discovered a comet that returns every 70 years that was named for him.
- 1826 The Olbers' paradox—why is the sky dark at night when the stars and galaxies are so bright? His answer to his own question is that the universe cannot be an infinite static arrangement of stars; this led to the theory of the expanding universe.

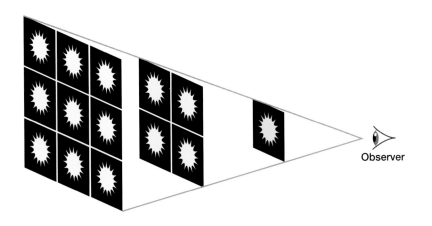

Observer

New Techniques and New Horizons in Astronomy

Above: American astronomer James Edward Keeler, director of the Allegheny observatory 1891–98.

Opposite: Giant reflector telescope in Dublin in 1881.

Below: A comet against a background of stars: Photographs like this replaced conventional star maps by the end of the century.

The second half of the century saw the observational basis of astronomy transformed by two new techniques of data analysis— photography and spectroscopy—while the ever-increasing power of telescopes revealed more and more about the stars and planets.

In telescope design there were persistent difficulties with both types of instrument, the refractor and the reflector. As the lenses of refractors grew larger and more powerful, the tubes had to grow to accommodate their focal length, and it became more difficult to maintain rigidity to prevent the tube from flexing under its own weight and so distorting the image. Between 1840 and 1890 the tubes of these refractors, almost 100 feet long, with lenses three feet in diameter, reached the physical limit of their development.

The alternative type of telescope, the reflector, was much shorter, but the principal difficulty was the mirror, which had to be cast from metal in a precise concave form and polished to gather and reflect the light. They were enormously heavy and difficult to cast, and they tarnished easily. The potential for much larger reflectors developed after the 1860s, when physicist Jean Foucault (1819–68) in Paris pioneered the technique of silvering glass mirrors. They were lighter, easier to cast and easier to polish than metal mirrors, and by the end of century, the reflector was undoubtedly the telescope of the future. (See also Volume 5, page 27.)

NEW SIGHTINGS

The improved power of the telescope yielded its clearest results in planetary astronomy—which could still reveal some surprises. As late as 1895 the rings of Saturn were proved to be neither solid nor liquid, but composed of meteorlike particles, when James Keeler made the necessary observations at the University of Pittsburgh Observatory. The composition of comets was analyzed, and it became evident that their heads did not merely reflect sunlight, but emitted their own radiant light, although how they did so was not clear. Perhaps the greatest interest centered on Mars. In 1877 it approached very close to the Earth, as it does about every 15 to 17 years. In that year Asaph Hall (1829–1907), astronomer at the U.S. Naval Observatory, was the first to observe two satellites, which he named Phobos and Deimos, in Greek "fear" and "panic"—the attendants of the war-god. They are extremely tiny, respectively only 25 and 15 kilometers (15.5 and 9.3 miles) in diameter, and Hall explained he was able to detect them because he had

Above and Opposite: Two striking early photographs of the Moon, taken by Warren de la Rue on a 13-inch reflecting telescope in the middle of the century (around 1858) at an observatory just outside London.

PERCIVAL LOWELL (1855–1916)
- Astronomer.
- Born in Boston, Massachusetts.
- Educated at Harvard.
- 1894 Established the Lowell Observatory in Flagstaff, Arizona.
- Fascinated with the planet Mars; created a series of maps showing lines on the planet's surface—he called them canals and suggested the possible existence of Martians.
- A prolific author of popular astronomy books, in 1906 he published *Mars and Its Canals*.
- 1907 Led an astronomical observation expedition to the Chilean Andes; returned with the first clear photographs of Mars.
- 1910 Published *Mars as the Abode of Life*.
- Postulated the existence of Planet X by observing the orbital "wobbles" of the planets Neptune and Uranus. Accurately predicted the brightness and position of the planet Pluto that was not discovered until 14 years after his death.

calculated from Mars's mass and gravity where their orbits might lie.

There is a curious history behind the story of Mars's satellites, for their existence had been predicted in the seventeenth century by Kepler. On the basis of a numerical progression from the Earth's one satellite to Jupiter's four (then known) he felt that Mars should have two. This idea became widely known, and their discovery was predicted by Jonathan Swift (1667–1745), the Anglo-Irish clergyman, in *Gulliver's Travels*. Mars became even more a focus of attention when, soon after the 1877 approach, the Italian astronomer Giovanni Schiaparelli (1835–1910) announced his claim that he had observed artificial canals on its surface. This claim was taken up in the 1890s by the American Percival Lowell (1855–1916), who supported the idea that intelligent beings on Mars had built the canals to irrigate the planet with melt-water from polar ice. The possibility of life on Mars continued to fascinate science writers for a century. Lowell spent many years trying to locate an unknown planet believed to lie beyond Neptune, which was finally discovered in 1930 and was named Pluto.

PRESERVED IMAGE

Even the most powerful telescopes could reveal little directly about the individual stars, but the application of photography to astronomy did have a huge effect on cosmology. Beginning in the 1850s, the Moon and Sun were naturally the first celestial objects to be photographed. The Sun's corona, seen during eclipses, attracted special interest because its immense glowing halo suggested that the Sun had an atmosphere. But it was when photographs were taken through powerful telescopes of star fields that its importance was really revealed, for a time-exposed photographic plate could detect light sources far fainter than the human eye, and it could preserve the image for future study.

In 1882 a Scottish astronomer at the Cape Observatory in South Africa, David Gill (1843–1914), made some excellent photographs of a bright comet then passing through the sky. But when the plate was exposed, no less striking than the comet were the number and clarity of stars in the background. It was as a result of photographs like these that a conference in Paris in 1887 decided to create a new series of celestial maps of unprecedented detail, not drawn by hand but in photographic form. The international *Carte du Ciel* (Map of the Heavens) project took many decades to complete, but it revolutionized the practice of stellar astronomy because these plates

could be studied by all scientists years after they were taken. The astronomer was no longer compelled to spend endless nights watching the sky, and special instruments were built to measure and coordinate the photographic plates. These time-exposed photographs required a clockwork mechanism to be geared to the telescope in order to hold a field of view, sometimes for an hour or more.

Photography was especially important in the study of the nebulas, in determining whether they were star fields or gas clouds. Henry Draper in America and Isaac Roberts in England made many historic early photographs of nebulas, some of which are, as we now know, distant galaxies in their own right.

COLORFUL UNIVERSE

The technique that had the most profound implications for astronomy and cosmology was undoubtedly spectroscopy, which analyzed the light from stars. Pioneered in the late 1850s by Bunsen and Kirchhoff (see above, page 37), within a decade the spectra of several thousand stars had been recorded and compared, and it became evident that they fell into groups. The Italian astronomer Angelo Secchi (1818–78) considered that there were four basic types, which ranged from those in which the white-blue light predominated to those in which the yellow-red end of the spectrum was most evident. These types could be explained simply as showing the different temperatures of the stars. The hotter the star, the more blue-white light; the cooler, the more yellow-red.

Sharp-eyed observers had long known that the stars were of different colors —Ptolemy had spoken of "golden-red Arcturus"— and now here was a clear scientific explanation. In the 1880s and 1890s a team of observers at Harvard College Observatory under American E.C. Pickering (1846–1919) examined thousands of spectra and expanded Secchi's four types into ten. The implication began to dawn on astronomers that these spectral types might represent the stages in the life-cycle of a star. Stars were perhaps evolving, cooling from white heat to dull-red heat, just as metals from a furnace do. And just like those metals, the process was revealed in their spectra, only in the case of stars it must occur over eons of time.

The implications of this insight, and of the photographs of nebulas and star-fields, would be fully drawn out only in the twentieth century, when they provided one of the foundation stones for a startling new vision of the age of the universe. Astronomy was still an observational and deductive science, for it could not handle and experiment with its subject matter in the way that chemistry or physics could. Yet new and unexpected links with chemistry, with photography, with new techniques in optics and engineering had enormously enlarged the data available to astronomers and had brought their science to the threshold of a new age.

Isaac Roberts, British pioneer of astro-photography.

Opposite: The Andromeda Galaxy, M31, taken by Isaac Roberts on December 29, 1888. It was taken on a twin telescope—a 20-inch reflector and a seven-inch refractor. It was taken with an exposure of four hours from a home observatory.

Foucault's Pendulum

Jean Foucault, who gave a visible proof that the Earth is rotating, three centuries after Copernicus had argued that it was.

A modern pendulum demonstrating the rotation of the Earth at the Science Museum, London, England, since 1996.

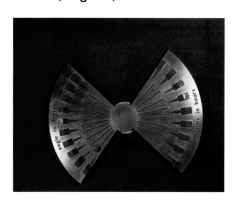

The nineteenth-century genius for devising fundamental experiments in the physical sciences is nowhere better seen than in the career of French physicist Jean Foucault (1819–68). A science journalist, Foucault never became a university teacher, and he pursued his experiments privately before announcing them. Foucault's most famous experiment was to give the first irrefutable proof that the Earth was rotating on its axis. He had been studying the motion of the pendulum for some time, and he noticed that moving the framework on which the pendulum was mounted did not appear to shift the plane in which it swung. This gave him his idea for a brilliant demonstration of the Earth's rotation.

In the cellar of his house he set up a pendulum with a five-kilogram (176 pound) bob suspended by a two-meter (6½ feet) thread, mounted so as to be free to swing in any direction. On January 8, 1851, Foucault set the pendulum swinging, and over the following hours he was delighted to observe that the path of its swing appeared to rotate around the room. Foucault was confirmed in his belief that the pendulum was swinging true, while it was the surface of the Earth that was rotating beneath it. Something that had been a scientific principle since the era of Copernicus had at last been demonstrated simply and clearly.

SCIENTIFIC SENSATION

A few weeks later the experiment was repeated with an 11-meter (36-foot) pendulum at the Paris Academy of Sciences and then again with an even longer pendulum before all the world at the Pantheon, the secular temple in Paris, where a 360° base-plate enabled the audience to watch the shift in its plane of swing. The experiment became something of a sensation that was widely reported in the popular press and imitated by scientific societies throughout the world. Foucault became famous and was awarded the Cross of the Legion of Honor. In fact, Foucault realized that his experiment would only work perfectly at the North or South Pole, where the axis of the pendulum would be perpendicular to the axis of the Earth. At the poles the rotation would take place in 24 hours, as it should. At other latitudes the time would vary: In Paris it took 34 hours, and on the equator it would take almost 48 hours. Continuing his experiments into the mechanics of the Earth's rotation, Foucault invented the gyroscope in 1852, which he showed gave an equally clear demonstration of the Earth's rotation, since it,

Foucault's pendulum created a public sensation in the Pantheon in Paris in 1880.

JEAN BERNARD FOUCAULT (1819–68)
- Physicist.
- Born in Paris, France.
- Started training to become a doctor but hated the sight of blood; changed to study experimental physics.
- 1845 With Armand Fizeau was the first to take a clear photograph of the Sun.
- Worked out the speed of light using the revolving mirror method.
- 1850 Showed that light travels more slowly through water than through air by taking the inverse relation between their speeds and their refractive indices.
- Proved the wave nature of light and that light speed diminishes in proportion to the index of refraction.
- Investigated the eddy current—the electric current induced in metal by a moving magnetic field—the Foucault current.
- 1851 Foucault's pendulum. He demonstrated the rotation of the Earth by using a freely suspended pendulum—demonstrated publlicly for the first time at the world's fair in the Pantheon in Paris.
- Constructed the first gyroscope in 1852.
- 1857 Made the Foucault prism.
- 1858 Improved the mirrors on the reflecting telescope—enabling the greater exploration of space.

too, maintained a fixed direction once it was revolving, even though its mounting was shifted. Some years later this discovery would lead to the invention of the gyrocompass for navigation, which gave a more reliable guide to true north than the magnetic compass needle.

Foucault's invention of the silver-glass mirror has already been mentioned, and he also experimented with the movement of light in various mediums. At first rather despised by the academic establishment, he was later acknowledged as an experimenter of genius. Foucault died of a brain tumor at the age of only 48.

Science and Technology

Above: Alexander Graham Bell, inventor of the telephone, from a photograph taken in 1876.

Below: Electric telegraphs as seen in an encyclopedia of the period. The electric telegraph began the modern revolution in communications.

The first Industrial Revolution began in England in 1760s and was associated with new techniques of iron and steel production and with steam power. It is generally agreed that this revolution had a craft basis: New solutions were found to old problems by talented engineers, but there was little or no input from theoretical science. From the 1780s onward, increasingly intricate machines were employed in industries such as textiles, but again they were technical improvements unrelated to pure science. Even the invention of gas lighting shortly before the year 1800 was not an achievement of pure science: It was found that coal heated (but not burned) to produce coke for furnaces gave off a gas that could be easily collected, stored, and burned in a controlled flame. First buildings then city streets were gas lighted by the 1830s.

By contrast, the years 1840 to 1880 witnessed a second Industrial Revolution that stemmed directly from the scientific advances outlined in this book. The effect of science on society in this period took many forms, but it can be most clearly seen in the advent of electrical power and in the chemical industries. Scientists of the caliber of Kelvin, Helmholtz, and Liebig involved themselves closely in these industries, applying science to the improvement, as they saw it, of social life.

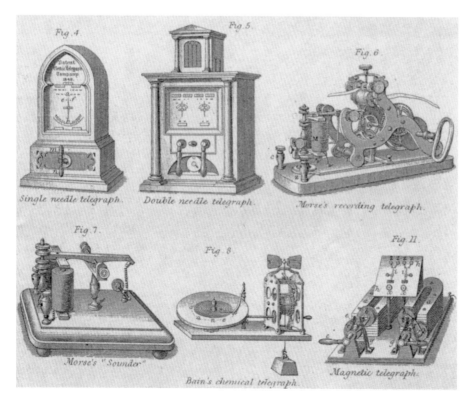

Single needle telegraph. Double needle telegraph. Morse's recording telegraph.

Morse's "Sounder" Bain's chemical telegraph. Magnetic telegraph.

SENDING SIGNALS

Among the very first applications of electricity was the telegraph, which arose from the realization that electrical impulses could be used to transmit signals along a wire. Oersted's discovery that an electric current deflected a magnetized needle led to experiments in the 1830s in which needles were made to point to letters and numerals and spell out messages. This system was used by railroad companies in England until it was swept away by the Morse code, in which pressing a single key completed a circuit and sent an impulse signal through a wire: It was named for American Samuel Morse (1791–1872). The first telegraph line using Morse code was opened in 1844 between Baltimore and Washington, and within another 20 years telegraph lines were laid on the Atlantic seabed, linking most of Europe and North America. For the first time in human history long-distance communication was possible at speeds faster than the horse or the ship could travel.

Thomas Edison, inventor of the electric light bulb and of the phonograph.

By 1880 the telegraph was overtaken by Scottish-born American Alexander Graham Bell's (1847–1922) telephone, in which the fluctuating air pressure caused by a speaker's voice was picked up by a diaphragm and turned into an electric current. It did not sweep away the telegraph, however, since many decades would be required to install the necessary equipment for transmitting the signals, while the telegraph structure was already in place.

BRINGING LIGHT

Once Faraday had demonstrated the principle of the electrical generator—that turning a coil of wire between the poles of a magnet induced a strong current in the wire—it was a comparatively short time before powerful and effective generators were made. But uses had to be found for this new source of power: It must be converted into mechanical energy by using it to turn electrical motors or into light or heat by creating an arc—a spark between two <u>electrodes</u>. One of the first uses of the powerful arc-light was in lighthouses from the 1850s onward. But many years of experiment were needed before a practical low-power light could be devised, which was the achievement of U.S. inventor Thomas Edison (1847–1931) and English chemist Joseph Swan (1828–1914) in 1880. With the invention of the filament light bulb a huge potential market for electricity was opened up.

The world's first power plant was opened in New York in 1882, and London, Paris, and other cities quickly followed. Lighting was required principally at night, so the operators of the power plants looked for other application, and the electric tramway was introduced in many cities. The cost of developing the infrastructure prevented electricity from being applied immediately to long-distance rail routes.

In this period famous companies with famous names were set up by the leading inventors. In addition to Edison and Swan there were the Siemen brothers (German, Ernst 1816–92 and adopted English,

THOMAS ALVA EDISON (1847–1931)
- Physicist.
- Born in Milan, Ohio.
- Partially deaf.
- Little formal teaching or training.
- Published *Grand Trunk Herald*, a newspaper for the Grand Trunk Railway.
- 1871 Invents the ticker-tape machine.
- 1876 Moves to Menlo Park, New Jersey. Becomes widely known as the "Wizard of Menlo Park" as he becomes a prolific inventor with over 1,000 patents.
- 1877 Invents the phonograph.
- 1879 Invents the electric light bulb.
- 1881–82 Designs first powerplant to distribute electricity.
- 1912 Produces first moving pictures.

Above: Sir Joseph Wilson Swan (1828–1914), who joined up with Edison to form the Edison and Swan United Electric Light Company.

Opposite, Above and Below: An ice-making machine of 1858—refrigeration brought a major improvement to food storage, particularly on board ships. The lower photograph shows the ice room of the SS *Victoria*.

Charles 1823–83), American George Westinghouse (1846–1914), and English Sebastian Ferranti (1864–1930), who all sought to promote new domestic and industrial applications of the power they were generating. The generators had of course to be turned, for which steam engines and later steam turbines were still used. In this sense the new electric power was still thermal energy produced by fire. But the point about electrical power was that it could be produced at a central point and conveyed to a remote site. Every user did not require his own steam engine to obtain power. Westinghouse also experimented with hydroelectric turbines, installing turbines under Niagara Falls.

CHEMICAL ACTION

The second great industry to be developed in the nineteenth century was chemicals. Bleaches and dyes were required in huge quantities by the textile industry, involving the mass production of acids and alkalis. Dyes in particular were complex organic chemicals for which synthetic substitutes had to be found, and this work could only be carried out by trained chemists. This was still more true of the complex form of glucose (sugar, composed of carbon, hydrogen, and oxygen), which was used to produce the first plastics, such as celluloid, and the first artificial fibers, such as rayon, developed in the late 1880s. It was also central to the new generation of high explosives such as nitroglycerine, which were used for blasting in the mining industry and for cutting long road and rail tunnels in the Alps and the Rockies.

Their use in weapons represented an early application of science to military technology, though in general nineteenth-century armies had found little use for new technology. Steam power was of no use on the battlefield, but high-explosive bullets and shells, combined with the rapid-fire guns, would revolutionize battle tactics.

The extraction of metals was also transformed by chemistry. Low-grade ores could have their impurities removed by heating in furnaces with suitable linings, leaving the pure metal to run free. A classic example is the process invented in the 1870s for removing phosphorus from low-grade iron by lining a furnace with alkaline limestone, with which the phosphorus combined.

Aluminum is one of the world's most common metals, but it does not occur in nature in a pure form. It must be extracted from bauxite. Tiny quantities were obtained in the laboratory, but it was only in the 1880s that electrolysis was found to release large amounts of the pure metal. Processes of this kind were not discovered by accident, but were the product of applying scientific principles. In addition to new basic materials such as plastics and light metals, industries producing pigments, adhesives, cleaners, and fertilizers all developed into major chemical enterprises.

COLD COMFORT

Of all the developments in food processing and handling, refrigeration was the most important and had a huge effect on the economies of Europe and the Americas. Commercial refrigeration began in the 1850s and is based on an understanding of the thermal exchange that occurs when certain liquids evaporate and then are recondensed.

In the first process they absorb heat from the environment, and in the second they emit heat. Refrigeration units were installed in railway cars and in ships, permitting meat to be imported into Europe from the Americas and Australasia. The economies of, for example, Argentina and New Zealand were entirely indebted to this technology.

By the end of the century the new sciences of electricity and chemistry were transforming the way people lived and worked, and the way they communicated, traveled, ate, and dressed. Even greater changes were near, with the internal combustion engine and the pharmaceutical industry. Science had become no longer a system of ideas, but an all-powerful force in society.

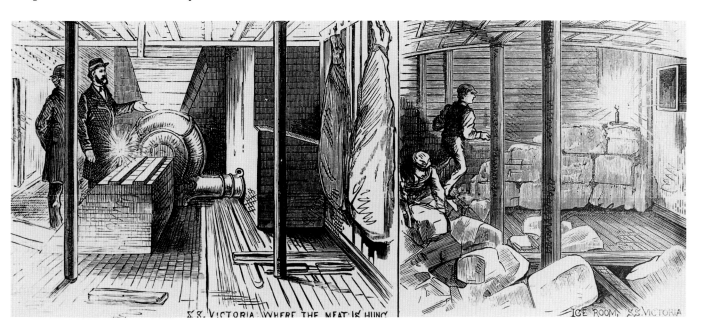

The Nineteenth-Century Revolution in Physical Science

If one had to characterize the achievement of the physicists and chemists of the nineteenth century, it could be summed up as the discovery of the invisible. By experiment and reasoning scientists had revealed hidden forces and patterns in nature whose very existence had been unsuspected by the leading minds of the previous century. Physicists had shown that any mechanism, whether it was a clanking steam engine in a small workshop or the universe of stars, functioned by changing energy from one form to another. The ultimate source and the ultimate end of this thermodynamic process could only be guessed at, but some fundamental secret had clearly been uncovered about the working of the universe.

SYMMETRY

The conservation of energy was complemented by the chemists' discovery of the conservation of matter through all the changes that occurred in chemical reactions. Metals, acids, and gases fused together or separated to create myriad compounds all from a few basic elementary substances. But throughout all this activity perfect equilibrium and symmetry were preserved. This symmetry was explained by the concept of the atom: the infinitely tiny particles that were the building bricks of matter, and that combined with each other according to invariable laws of number and proportion. It seemed as though the ancient belief in number mysticism—that mathematical structures were built into the fabric of the universe—was true after all. These atoms were purely hypothetical. Nobody had seen or examined them, but they represented a brilliant and logical answer to the puzzle of chemical reactions. It was also clear that the mysterious force of electricity was somehow involved in bonding the atoms of matter together, since an electrical charge could dissolve compounds into their elements.

When applied to the processes of life, chemistry tied the living and nonliving worlds together in another invisible network of relationships. Living tissue and living metabolism consisted, in the

The exhibition hall of a scientific institute in London, around 1850, displaying the technology of the era.

final analysis, of the same inorganic elements that made up the rest of the animate universe. Life had a purely physical basis, raising the baffling question of explaining the distinction between those metals in the Earth's crust or gases in the atmosphere and the forms into which they were organized in the bodies of plants, animals, and people. The exchange of these elements between living bodies and their environment was a perpetual process without which life of every kind would cease.

WIDER HORIZONS

If there was one field of science that had always been imagined to be forever out of reach of human knowledge, it was the nature of the stars. Yet within a few short years between 1850 and 1870 the techniques of spectroscopy and photography revealed that the stars, including our Sun, were composed of the very same chemical elements as the Earth and as the human body. Moreover, the light and energy radiating from the Sun were shown to be intimately connected to the electricity that bonded matter together, and whose power it was possible to unlock and use for humankind's own purposes.

None of these epoch-making discoveries was final; on the contrary, they seemed always to lead on to even deeper questions about how the various forces of nature were related to each other. But their profound nature made the nineteenth century truly a new age of science. Also new was the application of fundamental science to new technologies that would change human society in ways no one could possibly predict.

Top: Sir William Henry Perkin (seated at right), J. Dix Perkin, and laboratory assistants in 1870. Perkin (1838–1907) isolated the first synthetic dyestuff, mauve, produced from chemicals derived from coal tar and established the modern synthetic dyestuffs industry. In 1873 he synthesized coumarin, a sweet-smelling substance that initiated the synthetic perfume industry.

Above: The original bottle of mauveine dye as produced in 1856.

Glossary

Captions to illustrations on page 72.

Top: Edison's "Black Maria" studios in Orange, New Jersey. Edison was the United States' most prolific inventor with over a thousand patents to his name.

Bottom: Apparatus to prove Joule's Law (see page 7).

An unusual form for a factory—this looks more like an Egyptian temple with its fancy columns. The watercolor shows Marshall's flax mill in Rolbeck, Leeds, England, in the 1840s.

amp (ampere) the basic measure of the strength of an electric current.

angstrom the smallest unit of measurement in scientific use; one ten-millionth of a millimeter.

atom the smallest particle of an element that can exist either alone or in a combination.

combustion rapid chemical process that produces heat and light.

compound composed from the union of separate elements.

electrolysis the producing of chemical changes by passing an electric current through a nonmetallic conductor.

electrode a conductor that makes electrical contact with the nonmetallic part of a circuit.

electromagnetic spectrum the range of wavelengths of electromagnetic radiation from gamma rays to radio waves and including visible light, infrared, ultraviolet, and X-rays.

electron an elementary particle consisting of a charge of negative electricity.

element a fundamental substance consisting of only one kind of atom.

entropy the degree of disorder or uncertainty in a system; the process of running down.

fermentation breakdown of an energy-rich compound by enzymes.

gas a fluid with no shape or volume that expands indefinitely.

joule unit of energy equal to the work done by the force of one newton through a distance of one meter.

mechanics the transfer of energy from one place to another by means of physical components such as a lever.

molecule the smallest particle of a substance composed of one or more atoms.

organic chemistry chemistry involving carbon and its compounds that are fundamental to the processes of life.

proton an elementary particle identical to the nucleus of a hydrogen atom that, with neutrons, is a constituent of all other atomic nuclei and carries a positive charge.

reaction chemical change.

resistance the amount of electromotive force lost when it is carried through any medium. Resistance is what causes current electricity to become heat in an electric coil.

spectroscopy the investigation of the visible region of the electromagnetic spectrum.

telegraph a system for communicating at a distance by electric transmission via wire.

thermodynamics branch of physics that deals with the transformation of heat into work and vice versa.

valence relative capacity to unite, react, or interact.

volumetric analysis analysis relating to the measurement of volume.

The nineteenth century saw the industrialization of the world and "dark Satanic mills" built everywhere. They were usually near water both for its uses in industrial processes and also for the transportation facilities it offered. This mill is at Saltaire in Yorkshire, England, seen around 1869.

Captions to illustrations on page 73.

Top: Title page of Dalton's *A New System of Chemical Philosophy.*

Bottom: Bunsen's thermostat of 1869 that was used for determining the specific gravity of steam and gases.

71

PERIOD	1790	1800	1810	1820	1830	1840	1850

WORLD EVENTS

1790 Building of Washington D.C.
1800–15 Napoleonic Age in Europe
1803 Louisiana Purchase doubles the size of the U.S.
1807 Slave trade abolished in British Empire
1808–28 Independence movements in Spanish America

1840 British occupy Sind, Kashmir, Punjab and become masters of India
1845–48 U.S. takes Texas, New Mexico, and California
1848 Revolutions in many European cities
1848 Marx & Engels publish the Communist Manifesto

SCIENCE

1770 Lavoisier & Priestley both isolate oxygen
1789 Lavoisier's theory of elements published
1799–1803 Von Humboldt in South America
1800–20 Cuvier pioneers paleontology
1805 Davy uses electricity to isolate sodium
1808 Dalton's atomic theory
1820 Berzelius proposes chemical symbols
1820 Oersted shows relation between electricity & magnetism
1821 Faraday converts electricity into mechanical energy
1822 First photographs by Niepce in France
1824 First dinosaur fossils identified by Montell & Buckland
1830–33 Lyell's *Principles of Geology* published
1830–35 John Herschel charts nebulae
1838 Bessel measures stellar parallax
1844 Electric telegraphs using Morse code
1844 *Vestiges of Creation* published
1846 First operation with anesthetics
1846 Discovery of Neptune

ART & CULTURAL EVENTS

1790–1830 Romanticism in literature, art, and music: Goethe, Wordsworth, Byron, Beethoven, Schubert, Delacroix, Friedrich
1837 Opening of the Victorian Age in England
1840s Age of railway building in England, France, Germany, and United States
1840–60 First classics of American literature: Longfellow, Whitman, Hawthorne, Melville, Emerson
1840 onward Rise of the European novel: Dickens, Balzac, Dumas, Thackeray
1845–65 Criticism of industrial society: Carlyle, Ruskin, Thoreau

1850   1860   1870   1880   1890   1900

1854–56 Crimean War

1857 Indian Mutiny

1860 Beginning of Italian unification

1861–65 Civil War in North America

1867 Russia sells Alaska to the United States

1867 Dominion of Canada established

1869 Suez Canal opened

1869 First transcontinental railway in United States

1871 German Empire proclaimed

1898 Spanish-American War: U.S. takes Philippines

1899–1903 Boer Wars

1849–54 Theory of thermodynamics pioneered by Helmholtz, Clausius & Kelvin

1849 Fizeau measures the speed of light

1850 Liebig develops concept of carbon cycle

1850s Photography applied to astronomy

1850s Claud Bernard pioneers physiology

1850s Mendel's experiments in plant genetics

1850s Liebig's theories of human metabolism

1859 Bunsen and Kirchhoff apply spectroscopy to chemistry

1859 Publication of *The Origin of Species*

1862 Pasteur announces germ theory of disease

1865–85 Nash and Cope's dinosaur finds in the United States

1865 Lister introduces antiseptics

1870 Maxwell's theory of electromagnetic radiation

1870 Hertwig and Fol demonstrate mammalian fertilization

1872–77 Voyage of HMS *Challenger*

1880 Hertz sends first radio waves

1880s Harvard spectroscopic star analysis

1882 First electric power plant opens in New York

A

NEW SYSTEM

OF

CHEMICAL PHILOSOPHY.

PART II.

BY

JOHN DALTON.

Manchester:

Printed by Russell & Allen, Deansgate,
FOR
R. BICKERSTAFF, STRAND, LONDON.
1810.

1851 Great Exhibition opens in London

1850–65 Haussmann rebuilds Paris

1860–90 American painting: Winslow Homer

1860s Arts & Crafts movement: William Morris

1867 Marx publishes first part of *Das Kapital*

1875–90 Impressionist painting in France

1885 First automobiles in Germany

1885 First skyscrapers in Chicago and New York

1890s Realistic drama in Europe: Ibsen, Shaw, Chekhov, Strindberg

1892–93 World's Columbian Exposition in Chicago

Resources

FURTHER READING

There is a wealth of books published on the history of science, particularly biographies of great scientists. The following list includes many large works that contain many further resources.

Adams, F.D.: *The Birth and Development of the Geological Sciences*; Dover Publications, 1955.
Bowler, P.: *Evolution: The History of an Idea*; University of California Press, 1998.
Bowler, P.: *The Norton History of Environmental Sciences*; W.W. Norton & Co., 1993.
Boyer, C. & Merzbach, U.: *A History of Mathematics*; John Wiley & Sons Inc., 1989
Brock, W.H.: *The Fontana History of Chemistry*; Fontana Press, 1992.
Butterfield, H.: *The Origins of Modern Science*; Free Press, 1997.
Clagett, M.: *Greek Science in Antiquity*; Dover Publications, 2002.
Cohen, I.B.: *Album of Science: From Leonardo to Lavoisier*; Charles Scribner's Sons, 1980.
Cohen, I.B.: *The Birth of a New Physics*; W.W. Norton, 1985.
Crombie, A.C.: *Augustine to Galileo: The History of Science AD400–1650*; Dover Publications, 1996.
Crombie, A.C.: *Science, Art and Nature in Medieval and Modern Thought*; Hambledon, 1996.
Crosland, M.: *Historical Studies in the Language of Chemistry*; Heinemann Educational, 1962.
Eves, H.: *An Introduction to the History of Mathematics*; Thomson Learning, 1990.
Gillispie, C.C. (ed.): *Concise Dictionary of Scientific Biography*; Charles Scribner's Sons, 2000.
Gillispie, C.C.: *Genesis and Geology*; Harvard University Press, 1996.
Hallam, A.: *Great Geological Controversies*; Oxford University Press, 1983.
Ihde, A.J.: *The Development of Modern Chemistry*; Dover Publications, 1983.
Jaffé, B.: *Crucibles: The Story of Chemistry from Alchemy to Nuclear Fission*; Dover Publications, 1977.
Jungnickel, C. & McCormmach, R.: *Intellectual Mastery of Nature: Theoretical Physics from Ohm to Einstein*; University of Chicago Press, 1986.
Koyré, A.: *From the Closed World to the Infinite Universe*; The Johns Hopkins University Press, 1994.
Kuhn, T.: *The Copernican Revolution: Planetary Astronomy in the Development of Western Thought*; Harvard University Press, 1957.
Lindberg, D.C.: *The Beginnings of Western Science*; University of Chicago Press, 1992.
Porter, R. (Ed.): *The Cambridge Illustrated History of Medicine*; Cambridge University Press, 1996.
McKenzie, A.E.E.: *The Major Achievements of Science*; Iowa State Press, 1988.
Morton, A.G.: *A History of Botanical Science*; Academic Press, 1981.
Nasr, S.H.: *Islamic Science—An Illustrated Study*; London, 1976.
North, J.D.: *The Fontana History of Astronomy and Cosmology*; Fontana Press, 1992.
Olby, R. (et al.): *A Companion to the History of Modern Science*; Routledge, 1996.
Parry, M. (ed.): *Chambers Biographical Dictionary*; Chambers Harrap, 1997.
Porter, R.: *The Greatest Benefit to Mankind: a Medicinal History of Humanity from Antiquity to the Present*; HarperCollins, 1997.
Roberts, G.: *The Mirror of Alchemy*; British Library Publishing, 1995.

Ronan, C.A.: *The Cambridge Illustrated History of the World's Science*; Cambridge University Press, 1983.
Ronan, C.A.: *The Shorter Science and Civilisation in China*; Cambridge University Press, 1980.
Selin, H. (ed.): *Encyclopedia of the History of Science, Technology and Medicine in Non-Western Cultures*; Kluwer Academic Publishers, 1997.
Uglow, J.: *The Lunar Men*; Faber and Faber, 2002.
Van Helden, A.: *Measuring the Universe: Cosmic Dimensions from Aristarchus to Halley*; University of Chicago Press, 1985.
Walker, C.B.F. (ed.): *Astronomy before the Telescope*; British Museum Publications., 1997.
Whitfield, P.: *Landmarks in Western Science: From Prehistory to the Atomic Age*; The British Library, London, 1999.
Whitney, C.: *The Discovery of Our Galaxy*; Iowa State University Press, 1988.

THE INTERNET

Websites relating to the history of science break down into four types:

* Museum sites that offer some history and artifact photography. This is often the easiest way to visit international sites or those of states too far away to get to in person.
* College or other educational establishment sites that often provide online learning or study resources.
* General educational sites set up by enthusiasts (often teachers) and historians.
* Societies or clubs.

Examples of these three types of website include:

Museums
http://www.mhs.ox.ac.uk/
Museum of the History of Science, Oxford, England.
Housed in the world's oldest surviving purpose-built museum building, the Old Ashmolean.

http://www.mos.org/
Museum of Science, Boston.

http://www.msichicago.org/
Museum of Science and Industry, Chicago.

http://www.lanl.gov/museum
Bradbury Science Museum, a component of Los Alamos National Laboratory.

http://www.si.edu/history_and_culture/history_of_science_and_technology/
Smithsonian Institution site.

http://www.sciencemuseum.org.uk/
National Museum of Science and Industry, London, England.

http://galileo.imss.firenze.it/
Institute and Museum of the History of Science, Florence, Italy.

http://www.jsf.or.jp/index_e.html
Science Museum, Tokyo.

Colleges or institutions
http://sln.fi.edu/tfi/welcome.html
Franklin Institute with online learning resources and study units.

http://www.fas.harvard.edu/~hsdept/
Department of the History of Science of Harvard University.

http://www.hopkinsmedicine.org/graduateprograms/history_of_science/
Department of the History of Science, Medicine and Technology at Hopkins.

http://www.lib.lsu.edu/sci/chem/internet/history.html
Louisiana State University provides excellent history of science internet resources and links.

http://dibinst.mit.edu/
The Dibner Institute is an international center for advanced research in the history of science and technology and located on the campus of MIT.

http://www.mpiwg-berlin.mpg.de/ENGLHOME.HTM
Max Planck Institute for the History of Science

http://www.princeton.edu/~hos/
History of Science @ Princeton.

http://www.astro.uni-bonn.de/~pbrosche/hist_sci/hs_sciences.html
History of sciences from Bonn University, Germany, including indexes on the history of astronomy, chemistry, computing, geosciences, mathematics, physics, technology.

Educational sites
http://echo.gmu.edu/center/
ECHO—Exploring and Collecting History Online—provides a centralized guide for those looking for websites on the history of science and technology.

http://www.wsulibs.wsu.edu/hist-of-science/bib.html
Provides reference sources in the form of bibliographies and indexes.

http://dmoz.org/Society/History/By_Topic/Science/Engineering_and_Technology/
Open Directory Project providing bibliography and links.

http://orb.rhodes.edu/
ORB—the Online Reference Book—provides textbook sources for medieval studies on the web. It includes the Medieval Technology Pages—providing information on technological innovation and related subjects in western Europe—and Medieval Science Pages, a comprehensive page of links to medieval science and technology websites.

http://www.fordham.edu/halsall/science/sciencesbook.html
This page provides access to three major online resources, the Internet Ancient History, Medieval, and Modern History Sourcebooks.

http://www2.lib.udel.edu/subj/hsci/internet.htm
The University of Delaware Library provides an excellent guide to Internet resources.

Societies
www.hssonline.org
History of Science Society provides for its members the History of Science, Technology, and Medicine Database—an international bibliography for the history of science, technology, and medicine.

http://www.chstm.man.ac.uk/bshs/
British Society for the History of Science.

Set Index